T0251714

# C$_{60}$ Buckminsterfullerene

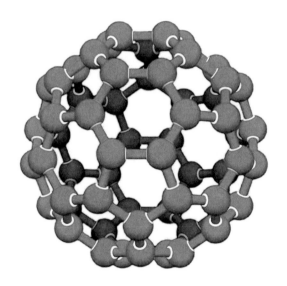

Sir Harry Kroto with giant fullerene structures. Photograph by Anne-Katrin Purkiss.

Pan Stanford Series on Nanomaterials and Nanotechnology  **Volume 1**

# C$_{60}$ Buckminsterfullerene
## Some Inside Stories

Edited by
# Harry Kroto

PAN STANFORD PUBLISHING

*Published by*

Pan Stanford Publishing Pte. Ltd.
Penthouse Level, Suntec Tower 3
8 Temasek Boulevard
Singapore 038988

Email: editorial@panstanford.com
Web: www.panstanford.com


**British Library Cataloguing-in-Publication Data**
A catalogue record for this book is available from the British Library.

ISBN 978-981-4463-71-3 (Hardcover)
ISBN 978-981-4463-72-0 (eBook)


Printed in the USA

# Contents

*Come in $C_{60}^+$, after nearly 100 years your number is finally up*    ix

### Part A

A   **Introduction**                                               3
*Harry Kroto*

1. Carbon Cluster Studies                                          7
2. Carbon Chains in Space and Stars                                9
3. Carbon in Space Dust and the Diffuse
   Interstellar Bands (DIBs)                                       10
4. $C_{60}$ Pre-discovery Experiments                              11
5. The Discovery of $C_{60}$                                       14
6. $C_{60}$ Theoretical Prehistory                                 15
7. The Circumstantial Supporting Evidence                          19
8. Skeptical Studies                                               22
9. Spectroscopic Theory                                            23
10. Extraction and Structure Proof                                 24
11. Aftermath Studies                                              27
12. Mechanism                                                      29
13. Epilogue: Detection of $C_{60}$ in Space                       30

### Part B

A1   **History of "the nozzle"**                                   39
*Lennard Wharton*

A2   **Up the carbon path**                                        47
*David E. H. Jones*

A3  In my time: scenes of scientific life (extract)    49
Sydney Leach

A4  Early days in the Rick Smalley lab    53
Michael A. Duncan

A5  Discovery of IRC+10216    69
Eric Becklin

A6  The discovery of the fullerenes    73
Sean C. O'Brien
April/May/June 1985: The Rotating Disc Source    75
August/September 1985    79
Metallofullerenes    81
Fullerenes    82
Photophysics of the Fullerenes    83
Conclusions    83

A7  How I conceived soccerball molecule C$_{60}$    87
Eiji Ōsawa
Encounters in Princeton    88
3D Aromaticity    89
Spherical Aromatic Carbon    90
Lucky Incident    90
Fullerene Fever    92
Epilogue    92

A7a  Partial translation of the book *Aromaticity* (in Japanese)    97
Zen-ichi Yoshida and Eiji Ōsawa

A8 The first stepwise chemical synthesis of $C_{60}$     **101**

*Lawrence T. Scott*

A9 $C_{60}$, Arizona, Don Huffman, and other stories     **113**

*Lowell D. Lamb*

1985     113
1986–1988     115
1989     118
1990     120
The Main Event     121
1991 to Present     125

A10 A PhD student's account of the $C_{60}$ story     **127**

*Jonathan Hare*

Flashback     128
The Scientific Background     130
Degree to PhD     131
A New Postgrad: The First Experiment     132
A Few Extracts from my Notebooks     135
The Red Solution     139
The Birth of Fullerene Science     140
Diary Entries from the Time     145
Harry Kroto's Nobel Prize, 9 October 1996     146
The Answerphone Recording     147

A11 The $C_{60}$ buckminsterfullerene formation process:
new revelations after 25 years     **149**

*Paul Dunk*

Starting Out     149
Research in a Foreign Country     151
Enter the Fullerenes     154

Building the Instrument                                    157
The Smallest Stable Fullerene (so far!)                    160
Closed Network Growth of Fullerenes                        163
What It All Means                                          174

**Part C**

**References**                                             **179**
*Index*                                                    181

# Come in $C_{60}^+$, after nearly 100 years your number is finally up

The Basel group of John Maier has made an outstanding breakthrough in unequivocally assigning two diffuse interstellar bands to the ion $C_{60}^+$ and so has succeeded in identifying a species in space which has eluded astronomers and other scientists for nearly 100 years.

In 1919 Heger published observations of some curious absorption lines in the spectra of stars. These lines were later shown to be due to material in the interstellar medium in the line of sight and not associated with the background star. The lines were broader than atomic lines (and so possibly molecular) and have since been called diffuse interstellar bands (DIBs). At the present time some 400 are known and they have been well documented observationally, and although scores of suggestions have been made as to the possible carriers, until now no single DIB has been identified. This is strange as the carrier or carriers must be stable in the interstellar medium and yet no terrestrial species has been found to correlate with any observed band. A consensus has grown that the carriers are probably molecular species, although some favoured absorption by atoms in grains.

At last after nearly 100 years the group of John Maier in Basel has made the first breakthrough by assigning two DIBs unequivocally to $C_{60}^+$, the positive ion of $C_{60}$. This remarkable achievement was no accident as it was the result of many years of painstaking work developing state-of-the-art spectroscopic techniques to create conditions in the laboratory which simulated the extremely low temperatures and low pressures in the interstellar medium. The $C_{60}$ molecule buckminsterfullerene was serendipitously discovered in our experiments in 1985 which had as their second aim the identification of the DIBs. The fact that this molecule was spontaneously created under conditions in which possible contenders for DIB carriers might also be produced seemed like an intriguing coincidence worth following up. In 1987 I suggested that if $C_{60}$ was a likely DIB contender in interstellar space, then the ambient radiation field would almost certainly ensure that it must be ionised as $C_{60}^+$. Maier in 1993 measured the spectrum of the ion in a matrix and found two strong lines, the frequencies of which were used by Foing and Ehrenfreund to search for new DIBs close to these frequencies. They found two lines in reasonable correspondence; however, significant matrix shifts meant that for an unequivocal assignment the spectrum had to be measured in the gas phase at a low temperature—a daunting task. Maier has spent much of the intervening time developing superb state-of-the-art experimental techniques to achieve just this, and this brilliant breakthrough is a result of his determined approach to solving one of the most important and long-lasting puzzles in science. It is a bit sad of course that this puzzle has now been solved at least in part; however, there are several hundred lines still to identify, but it now seems highly likely that other $C_{60}$ analogues are also present in the ISM and one might conjecture many of the other lines are due to these analogues which could be endohedral species in which an atom of, say, sodium or calcium is trapped inside the fullerene cage or the atom is attached to the outside of the cage. Such species should display quite strong spectra and be detectable in the future, although laboratory measurement will still be a daunting task.

It is incredible to think that one of the most abundant set of species in interstellar space may be these carbon cages and yet they are almost non-existent in the terrestrial environment and took until nearly the end of the 20th century to be discovered. The molecule was under our noses all the time in flames (indeed it is now made in bulk by combustion of methane) and at the same time its signature was being recorded whenever astronomers observed stellar spectra.

## 1. The Spectrum and the Fit to the DIBs

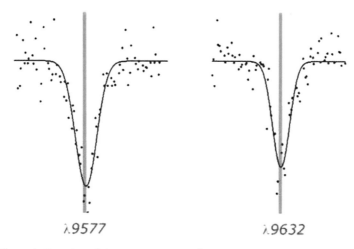

$\lambda 9577$ $\qquad$ $\lambda 9632$

**Figure 1** Gas-phase laboratory spectra of C$_{60}^{+}$ at 5.8 K. Gaussian fits to the experimental data (circles) are represented by the solid blue lines. The vertical pink lines are the rest wavelengths, 9,577.4±0.2 Å and 9,632.6±0.2 Å, of two reported DIBs.

## Laboratory confirmation of C$_{60}^{+}$ as the carrier of two diffuse interstellar bands

E. K. Campbell, M.Holz, D. Gerlich and J. P. Maier

The diffuse interstellar bands are absorption lines seen towards reddened stars. None of the molecules responsible for these bands have been conclusively identified. Two bands at 9,632 Å and 9,577 Å were reported in 1994, and were suggested to arise from $C_{60}^+$ molecules, on the basis of the proximity of these wavelengths to the absorption bands of $C_{60}^+$ measured in a neon matrix. Confirmation of this assignment requires the gas-phase spectrum of $C_{60}^+$. Here we report laboratory spectroscopy of $C_{60}^+$ in the gas phase, cooled to 5.8 K. The absorption spectrum has maxima at 9,632.7±0.1 Å and 9,577.5±0.1 Å and the full widths at half-maximum of these bands are 2.2±0.2 Å and 2.5±0.2 Å, respectively. We conclude that we have positively identified the diffuse interstellar bands at 9,632 Å and 9,577 Å as arising from $C_{60}^+$ in the interstellar medium.

### See also

Palca, J. "Buckyballs" solve century-old mystery about interstellar space. National Public Radio website. July 16, 2015. http://www.npr.org/sections/thetwo-way/2015/07/16/423291426/buckyballs-solve-century-old-mystery-about-interstellar-space?utm_source=facebook.com&utm_medium=social&utm_campaign=npr&utm_term=nprnews&utm_content=20150716

# Part A

# Introduction

## Harry Kroto

Department of Chemistry & Biochemistry, Florida State University,
95 Chieftan Way, Tallahassee, FL 32306-4390, USA
kroto@chem.fsu.edu

The story of discovery of $C_{60}$ buckminsterfullerene is not only multifaceted but also convoluted, and this makes the assembly of a coherent, straightforward account very difficult, if not impossible. The bare bones are to be found in three key papers: (a) An experiment, in 1985, designed to re-create the conditions in cool carbon stars serendipitously uncovered the existence an all-carbon species which was conjectured to be a molecule consisting of 60 equivalent C atoms located at the corners of a truncated icosahedral closed cage.[1*] (b) In 1990 the molecule was extracted by sublimation from the deposit of a carbon arc and the spheroidal shape proven by X-ray analysis.[2*] (c) Simultaneously, in 1990, the molecule was created independently, solvent-extracted

---

Note: [*] indicates a paper or article republished in here and A indicates an account specially written for this book.

---

$C_{60}$ Buckminsterfullerene: Some Inside Stories
Edited by Harry Kroto
Copyright © 2015 Pan Stanford Publishing Pte. Ltd.
ISBN 978-981-4463-71-3 (Hardcover), 978-981-4463-72-0 (eBook)
www.panstanford.com

and the truncated icosahedral pattern confirmed by the detection of a single NMR line and the chromatographic method of isolating fullerenes developed.[3*] Furthermore in this study the ultimate unequivocal confirmation that a whole family of fullerenes existed was obtained by the detection of the 5-line NMR spectrum for $C_{70}$. Then the field exploded and today, on average, about a thousand papers are published each year describing research advances involving fullerenes. An essentially complete review[4*] surveyed the status of the fullerene field up to 31 December 1990. It is arguable, and I would argue it, that the main aspect of the discovery was not the fact that $C_{60}$ could be created, but that it self-assembled spontaneously, because this resulted in a reassessment of our perspective on the general dynamic factors which control structure assembly processes at nanoscale dimensions. In so doing it kick-started nanoscience and nanotechnology and became an iconic character of the field.

By themselves the published research papers present a rather arid account, yielding little insight into how and why the breakthrough actually occurred. This is almost always the case in science where personal accounts are rarely available. However, in this case many were genuinely interested in the breakthrough and several personal accounts, and general articles were written by some members of the groups primarily involved in the discovery and extraction.[5-16] Numerous articles were published by journalists such as that by Taubes[17*] in general science magazines as well as two books[18,19] and a collection of rather intimate interviews by Hargittai.[20] The origins of the name "buckminsterfullerene" and the family name "fullerene" are discussed by Applewhite[21] and Nickon and Silversmith.[22] There was also a film produced by the BBC, some excerpts of which may occasionally be viewed on YouTube.[23] There are numerous indirect accounts often propagating inaccuracies, to be found littering the literature; for instance, several accounts claim, erroneously, that $C_{60}$ was first detected in space! Having perused the plethora of material available, I still feel that many interesting aspects are missing and this compendium is an attempt to fill some of the interesting gaps and help to create a more satisfying and intimate picture of how the discovery actually came about. The articles and reprints collected here should be considered more as a supplement to previous accounts offering a more detailed perspective on the breakthrough and the way advances in general often arise by the confluence of ideas and factors from widely differing sources.

When a convoluted event occurs, "Rashomon" factors occasionally apply in that some contributions do not appear to fit well together. The film *Rashomon* by director Akira Kurosawa, which is based on two short stories, "In the Grove" and "Rashomon" by Ryūnosuke Akutagawa, deals with this sort of situation. The determination of what actually happened was not Kurosawa's aim in this film as all the accounts are completely incompatible and embellished by each individual's personal ego. However, in a real situation something definite actually happens and the lesson I *personally* draw from *Rashomon* is that *in a real case* individuals tend to present their personal "realities" and these can offer deeper and more complete insights into multifaceted events. Basically Objectivity with a capital O is to be found in the Totality of the Subjectivity with capitals T and S, respectively. An interesting corollary of Kurosawa's thesis is that individuals, not directly involved who seek to present the definitive story, can also not be relied upon to evince disinterested accounts as they present their own personal "biased" views! The conclusion is that there is no such thing as a single "objective" story containing no conflicting elements!

In this compendium are to be found copies of most of the papers detailing advances that I personally (!) feel were crucial steps on the way to the discovery of the fullerenes and their extraction. Because existing accounts in my opinion[5-16] paint an incomplete picture, I have asked some key researchers who laid the foundations for the discovery or were involved with key aspects of the story to contribute to an introductory section of this book. Fortunately almost all have kindly agreed to pen their recollections. These new accounts provide fascinating new insights into how the breakthrough occurred and highlight the fact that scientific research is an intrinsically human and totally unpredictable activity. With hindsight I think one can recognize at least three major trails leading inexorably to the revelation that, like Orson Welles who lurked in the shadowy backstreets of Vienna in the famous film *The*

**Table 1** Diagramatic representation of the various disparate pathways which led up to the discovery of $C_{60}$ in the laboratory and space

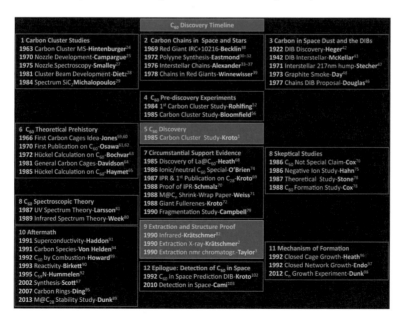

| $C_{60}$ Discovery Timeline | | |
|---|---|---|
| **1 Carbon Cluster Studies**<br>1963 Carbon Cluster MS-**Hintenburger**[24]<br>1970 Nozzle Development-**Campargue**[25]<br>1975 Nozzle Spectroscopy-**Smalley**[27]<br>1981 Cluster Beam Development-**Dietz**[28]<br>1984 Spectrum SiC$_n$**Michalopoulos**[29] | **2 Carbon Chains in Space and Stars**<br>1969 Red Giant IRC+10216-**Becklin**[38]<br>1972 Polyyne Synthesis-**Eastmond**[30-32]<br>1976 Interstellar Chains-**Alexander**[33-37]<br>1978 Chains in Red Giants-**Winnewisser**[39] | **3 Carbon in Space Dust and the DIBs**<br>1922 DIB Discovery-**Heger**[42]<br>1942 DIB Interstellar-**McKellar**[43]<br>1971 Interstellar 217nm hump-**Stecher**[47]<br>1973 Graphite Smoke-**Day**[48]<br>1977 Chains DIB Proposal-**Douglas**[46] |
| | **4 $C_{60}$ Pre-discovery Experiments**<br>1984 1st Carbon Cluster Study-**Rohlfing**[52]<br>1985 Carbon Cluster Study-**Bloomfield**[56] | |
| **6 $C_{60}$ Theoretical Prehistory**<br>1966 First Carbon Cages Idea-**Jones**[59,60]<br>1970 First Publication on $C_{60}$-**Osawa**[61,62]<br>1972 Hückel Calculation on $C_{60}$-**Bochvar**[63]<br>1981 General Carbon Cages-**Davidson**[64]<br>1985 Hückel Calculation on $C_{60}$-**Haymet**[65] | **5 $C_{60}$ Discovery**<br>1985 Carbon Cluster Study-**Kroto**[1] | |
| | **7 Circumstantial Support Evidence**<br>1985 Discovery of La@$C_{60}$-**Heath**[68]<br>1986 Ionic/neutral $C_{60}$ Special-**O'Brien**[74]<br>1987 IPR & 1st Publication on $C_{28}$-**Kroto**[69]<br>1988 Proof of IPR-**Schmalz**[70]<br>1988 M@C$_n$ Shrink-Wrap Paper-**Weiss**[71]<br>1988 Giant Fullerenes-**Kroto**[72]<br>1990 Fragmentation Study-**Campbell**[79] | **8 Skeptical Studies**<br>1986 $C_{60}$ Not Special Claim-**Cox**[76]<br>1986 Negative Ion Study-**Hahn**[75]<br>1987 Theoretical Study-**Stone**[78]<br>1988 $C_{60}$ Formation Study-**Cox**[78] |
| **8 $C_{60}$ Spectroscopic Theory**<br>1987 UV Spectrum Theory-**Larsson**[81]<br>1989 Infrared Spectrum Theory-**Week**[80] | | |
| **10 Aftermath**<br>1991 Superconductivity-**Haddon**[91]<br>1991 Carbon Species-**Von Helden**[94]<br>1992 $C_{60}$ by Combustion-**Howard**[99]<br>1993 Reactivity-**Birkett**[93]<br>1995 C$_{59}$N-**Hummelen**[92]<br>2002 Synthesis-**Scott**[67]<br>2007 Carbon Rings-**Ding**[95]<br>2013 M@C$_{28}$ Stability Study-**Dunk**[89] | **9 Extraction and Structure Proof**<br>1990 Infrared-**Krätschmer**[82]<br>1990 Extraction X-ray-**Krätschmer**[2]<br>1990 Extraction nmr chromatogr.-**Taylor**[3]<br><br>**12 Epilogue: Detection of $C_{60}$ in Space**<br>1992 $C_{60}$ in Space Prediction DIB-**Kroto**[102]<br>2010 Detection in Space-**Cami**[103] | **11 Mechanism of Formation**<br>1992 Closed Cage Growth-**Heath**[96]<br>1992 Closed Network Growth-**Endo**[97]<br>2012 C$_n$ Growth Experiment-**Dunk**[98] |

*Third Man*, a third well-defined allotropic form of carbon has, since time immemorial, been lurking in the dark recesses of the universe. It is hard to credit the fact that the discovery of a molecular allotrope of carbon did not occur until nearly the end of the 20th century when the element involved is the most multitalented and by far the most well-studied in the Periodic Table. The breakthrough involved the amalgamation of a kaleidoscope of disparate research studies and the diagram in **Table 1** has been devised to provide a semblance of rational order. Discoveries that appear to arrive from "left field" litter the sciences and serve as a ubiquitously unheeded warning to those who think they know how science should be done and what science should be funded.

## 1. Carbon Cluster Studies

The Carbon Cluster Pathway involved the development of techniques to study the mass spectra of molecules and clusters of refractory materials, in particular carbon. Between 1958 and 1963 Hintenberger and colleagues published a series of fascinating papers which contain what I consider to be landmark mass spectrometric observations on pure carbon species which they found in the products of arc-discharged graphite.[24*] They observed species with as many 33 carbon atoms! It is interesting to conjecture how different the history of the fullerenes might have been had their study extended to twice as many atoms! Another key event was the development of the supersonic nozzle by Roger Campargue which produced cold ensembles of molecules in the gas phase.[25] A review article outlining his personal perspective is reproduced here.[26*] The next step on this trail was the combination of the supersonic nozzle with a tunable laser in a landmark advance by

Rick Smalley, Lennard Wharton and Donald Levy which enabled them to produce molecules in the gas phase at sufficiently low internal temperatures which facilitated the analysis of extremely complex electronic spectra, such as that of $NO_2$.[27*] Lennard Wharton has written an account of his personal recollections (A1) which reveals just how much careful background work was involved in developing this major breakthrough. During the research on this introductory chapter, quite fortuitously, Sydney Leach, a long-time friend and occasional coworker, mentioned his involvement with the start of Roger's nozzle studies and Sydney has sent a short (already published) anecdote (A2) about the way in which Roger's original gas phase dynamics research programme came into being. This is a further fascinating insight into the seemingly haphazard and totally unpredictable way that important scientific breakthroughs occur.

The next crucial step and certainly the most technically crucial step was the creation of the laser vapourisation supersonic cluster beam instrument by Rick Smalley's group at Rice University in 1981.[28*] This advance enabled the mass spectra of large clusters created from refractory precursor materials to be detected for the first time and in some case the measurement of their spectra, such as $SiC_2$.[29] This elegant technique, more than any other, has revolutionised the study of refractory clusters, and Michael Duncan, who was a PhD student on this development, has also contributed a detailed and quite fascinating personal account (A3) which shows how this major breakthrough came about. In particular Michael describes how the machine, affectionately named "Ap2" which uncovered the existence of $C_{60}$, came to be constructed in the first place. It is my view that the discovery of $C_{60}$ was the raison d'être of this brilliant technical advance.

## 2. Carbon Chains in Space and Stars

The Carbon Chain Pathway started in a totally different field with the development of organic synthetic techniques by David Walton at Sussex who created extended linear carbon chain structures called polyynes with alternating single and triple bonds[30–32] and the study of the molecular dynamics of these chains by molecular spectroscopy (microwave rotational spectroscopy) in a collaboration between David and my spectroscopy group. In fact the key catalyst was a unique Chemistry by Thesis course, initiated by the then dean of the School of Molecular Sciences, Colin Eaborn. In this course chemistry undergraduates at the University of Sussex were able to obtain BSc degrees by carrying out research more or less full-time for ca. two years. The student involved in the study of the first cyanopolyyne $HC_5N$, Anthony Alexander, did an outstanding job.[33] When the rotational frequencies had been measured they were used in a collaboration between our Sussex group and Takeshi Oka, Lorne Avery, Norm Broten and John Macleod at the National Research Council (NRC) in Canada, and this resulted in the discovery by radioastronomy of the unexpectedly high abundance of $HC_5N$ in the interstellar medium.[34*] Then Colin Kirby, at the time a grad student, achieved the difficult synthesis of $HC_7N$ devised by David and measured its spectrum[35] which enabled us to detect it in space.[36] Then using the frequencies Takeshi imaginatively predicted the frequency of $HC_9N$ and we detected it as well.[37] It was this series of Sussex/NRC laboratory and radioastronomy studies which uncovered the, at the time amazing, abundance of the long carbon chain molecules in space. The next step in the story was the detection by Eric Becklin, Gerry Neugebauer and their coworkers of an amazing object emitting infrared radiation an order of magnitude greater than any previously observed IR source and the identification of the object as the cool red giant carbon star IRC+10216.[38] Eric has also sent a personal account (**A4**) which nicely captures the euphoria that often accompanies a moment when an important discovery is made, in this case of an exceptional new source of infrared radiation. A subsequent radioastronomy study focused on this star resulted in the exciting (certainly to me) observation by Gisbert Winnewisser and Malcolm Walmsley[39] that our long carbon chain cyanopolyynes molecules were being ejected from IRC+10216 into the interstellar medium. These results catalysed preliminary conjectures about the importance of carbon in the interstellar medium and stars[40,41] which were later to stimulate the experiments which resulted in the discovery of $C_{60}$.

### 3. Carbon in Space Dust and the Diffuse Interstellar Bands (DIBs)

A third primary pathway involves a couple of "Interstellar Mysteries" which are arguably two of the most intriguing in the whole of the sciences. One involves a discovery, made originally in 1922 by Heger, of some curious absorption features in the spectra of stars[42] which came to be known as the diffuse interstellar bands (DIBs). The interstellar nature of these features was unequivocally confirmed during the 1930s and their properties summarised by McKellar in 1940.[43] The DIB field has been extensively and carefully reviewed by Herbig.[44,45] The tantalising puzzle of the nature of the carrier is still today, almost a century later, unsolved, although scores of spectroscopists and astrophysicists have ruminated long and hard over the identity, observationally, experimentally and theoretically. The observing sessions which revealed the abundance of the carbon chains in space were carried out at the National Research Council Telescope in Algonquin Park Canada, and I went there a few times to participate in our observations. During one of these visits Alec Douglas discussed his thoughts with me on the carriers of the DIBs and in particular he suggested the possibility that extended carbon chain species related to those we had discovered might be involved.[46] A second important interstellar feature is also involved in the story. This is a strong absorption band at 217 nm which Stecher conjectured might be due to carbonaceous dust particles.[47] Subsequently Day and Donald Huffman carried out a very important laboratory study on the UV spectrum of carbon smoke and showed that this conjecture was quite convincing.[48*]

## 4. C$_{60}$ Pre-discovery Experiments

The specific aim of the discovery experiment was actually very simple: To simulate the supposed chemistry in the atmosphere of a star such as IRC+10216 and show that the polyynes HC$_n$N ($n$ = 5, 7, 9), which we had detected in the ISM, could be created when carbon atoms nucleate in an atmosphere containing nitrogen and hydrogen (e.g. NH$_3$) and so support my hypothesis that the chains originated in stars.[40,41] At the time, in situ ion molecule reactions[49,50] and/or grain surface catalysis[51] were the two strongest candidates able to account for most other interstellar molecules. In my mind neither theory could account for the chains we had observed as there appeared to be too many "heavy" carbon atoms to be created by the very slow processes in the very tenuous interstellar gas and they also were surely just too heavy to be sufficiently easily vapourised from a solid surface. Bob Curl had invited me to visit Rice after a conference organised by Jim Boggs at Austin, and during this visit in Easter 1984 Bob showed me a manuscript detailing a resonant 2-photon ionisation (R2PI) spectrum obtained by Rick Smalley's group which indicated that SiC$_2$ had, unexpectedly, a triangular structure.[29] This was extremely interesting to me as my group had focused over the previous decade on the creation of whole families of new molecules involving multiple bonds to second and third row elements (i.e. containing $>C=S$, $>C=Se$, $-B=S$, $>C=P-$ and $-C\equiv P$ moieties)[41] and I had been ruminating for quite a while over how we might tackle the problem of creating molecules containing the $>C=Si<$ moiety, which I knew would be quite difficult. Bob encouraged me to go over to see Rick in his laboratory and while Rick was describing Ap2, which was his pride and joy at the time, the basic experiment described above formed in my mind. Furthermore, the possibility of a second experiment also formed: This was to measure the R2PI spectra of various carbon chain species and confirm or otherwise the contention of Alec Douglas's that carbon chains might be the carriers of the DIBs.[46] That evening in the Curl household, I discussed these ideas with Bob, and over the following 17 months the technical issues, mainly pertaining to the difficulties involved in carrying out the more complex R2PI/DIB experiment, were the subject of letters sent to and fro between us (these were pre-email days!).

Two or three months after my Easter 1984 visit to Rice, I was sitting in the Sussex University Chemistry Laboratory coffee room when my colleague Tony Stace handed me a copy of a fascinating paper[52*] by a group at Exxon which to my amazement, and "slight" irritation, presented details of essentially the basic experiment I had proposed three months previously to the Rice group. This paper contained the most fascinating new observation that the overall carbon mass spectrum pattern was actually bimodal and that in addition to the set detected by Hintenberger's group,[24] which seemed to peter out around $C_{30}$ or so, a second set of only even numbered species began to appear above ca. $C_{30}$ rising to a maximum in the $C_{50}$ to $C_{70}$ region and tailing off around $C_{100}$. The Exxon group suggested these new features were due to "carbyne",[53] a mythical creature whose existence has been the subject of controversy for getting on for half a century.[54] In 1982 Smith and Buseck had shown essentially conclusively that the "carbyne" proposal had been based on an experimental artefact.[55] One thing about which I was pretty certain was that this hypothetical material made no more sense from a chemical point of view than the Loch Ness Monster makes from an evidence-based logical viewpoint. Polyynes explode with great violence if any attempt is made to isolate and condense them, and in our Sussex microwave spectroscopy work we took great precautions to ensure they remained in the vapour phase at low pressures. After the discovery experiment in September 1985[1] we realised that the mass spectrometric peak of $C_{60}$ in this Exxon paper was actually somewhat stronger than other adjacent peaks but the significance of this strength had gone unrecognised. A second paper by Bloomfield et al., also carried out prior to our experiment, had probed the photodissociation characteristics of $C_{60}$.[56] As far as I can tell, neither of these two groups varied the clustering conditions with a view to probing the creation process itself. Had this been done I suspect that both groups would have realized that the $C_{60}$ signal could be made dominant and this would have alerted them to the fact that something exceptional was involved and worthy of further careful examination. Thus we realised that $C_{60}$ had been in the literature for some 18 months prior to our discovery in September 1985. Although scores of researchers must have seen the peak (I also)—it was even labelled $C_{60}$—no one appears to have considered it seriously! When I first saw the Exxon paper my first reaction was that the new family of only even-numbered carbon species might be planar graphene flakes of various shapes and sizes. That even-numbered graphene flakes might be more stable than odd-numbered ones did not seem that implausible at the time.

## 5. The Discovery of $C_{60}$

In September 1985, about 17 months after my first visit, Curl and Smalley finally agreed that we should carry out the basic experiment which I had proposed. The carrot for the Rice group was that the basic experiment, to simply simulate a stellar carbon atmosphere, was, to all intents and purposes, a necessary preliminary to the much more complicated and seemingly "more important (!)" DIB study. The actual discovery experiments were carried out by the students Jim Heath, Sean O'Brien, Yuan Liu and me in the space of less than two weeks. A fourth student, Qing-Ling Zhang, who had only just started was also involved in some of these experiments. The students threw themselves wholeheartedly into the project, especially Jim Heath. We also had an improved weapon at our disposal, in the shape of a refined nozzle assembly, which Sean had designed. At the start of the experiments we had a rather irritating

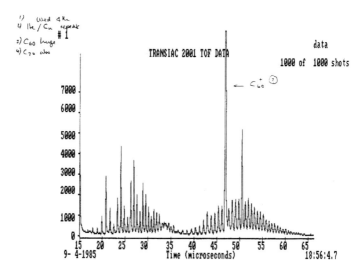

**Figure 1** Birthday "card" printout on 4 September 1985, the day on which $C_{60}$ was recognised as very special. On that day Yuan also noted that $C_{60}$ and $C_{70}$ are very strong in the Ap2 lab book. Reproduced with permission from Ref. 8. Copyright Wiley-VCH Verlag GmbH & Co. KGaA.

problem in that we were unable to print out our data, but fortunately Yuan worked hard to resolve this problem and within a couple of days, on 4 September 1985, we had a paper copy showing a dominant mass spectrometric peak at 720 amu which indicated that a species with 60 carbon atoms was very special indeed **(Fig. 1)**.

Although several personal perspectives on the discovery experiments are to be found in the accounts written by Bob Curl,[5,7,9] Rick Smalley[5,7,11] and me[6,8,10] the personal account that Sean has written for this monograph **(A5)** presents a valuable new perspective. It gives an intimate insight into the hectic discovery period in a way that only a young student (as Sean was at the time) can see it. Among other things, Sean describes how some key technical refinements that he made to the nozzle of Ap2 facilitated variation of the clustering conditions and enabled us to find a way of making the $C_{60}$ peak a dominant character among a veritable myriad of lesser ones. On 13 September, some nine days after the Buckyball's birthday on 4 September 1985 **(Fig. 2)**, the paper was received by *Nature*! The paper was accepted on 18 October, and it is interesting to peruse the comments of the referees**(Fig. 2)**.

The experiments which were the original aim, to use the cluster beam apparatus to simulate circumstellar shell chemistry, were actually carried out[57,58] and confirmed beautifully the conjecture that long cyanopolyyne chains may be readily created in the hot, chaotic, carbonaceous gas of red giant stars. Understandably, however, these studies were somewhat sidelined due to the antics of the newly discovered all-carbon prima donna.

## 6. $C_{60}$ Theoretical Prehistory

After news of our observations became known we quickly discovered that there had also been another fascinating

I have had a chance to read the very stimulating paper
on $C_{60}$: Buckminsterfullerene by H. W. Kroto, et al. It is
certainly an extremely energetic paper full of ideas and specu-
lations; in the spirit of stimulating scientific debate it
certainly is a fun paper. In terms of substantial content I am
not sure exactly what it contains other than the ability to
emphasize the 'production" of the specific carbon cluster size,
$C_{60}$. One needs to be careful about this "production" because the
process involved in detecting the species are complex and it is
possible to mistake ion production with neutral species concentration. The major difference between the approach to produce
these clusters compared to those previously discussed is the
higher helium pressure, and this can lead to the formation of
much colder species. This could significantly impact the
apparent ion production and give the impression of higher neutral
concentration. Without more experiments or more discussion of
experiments that the authors may have done it is impossible to
prove the validity of the claim that $C_{60}$ and possibly $C_{70}$ are
"magic number" clusters. There is no doubt that if the results
reported are correct and this approach has lead to a preferential
production of these specific materials, and if these approaches
can be scaled to produce "carbosoccrene", it would certainly open
up exciting new measurement opportunities, indeed new understanding of the chemistry and physics irrelevant to terestial as
well as astrophysics chemistry.

I think it would be useful if the authors were to refer
to some earlier work published in Nature by A. Douglas, 269, 130,
1977 on the carriers of the difused interestellar absorption.
Also, while I cannot give the exact reference, the publication by
W. Kratschmer, N. Sorg and D. R. Huffman.

*Referee B*

Comments on the manuscript by H.W.Kroto et al

Preferred (stable) numbers of atoms (and molecules) are not
unusual and indeed are well known. However, the observation
that the $C_{60}$ structure becomes so very dominant under certain
conditions is very interesting and should be reported. Other
than this, the Letter is highly speculative, but much of he
speculation is very interesting. However, the statement on P.3.
"its stability when formed under the most violent conditions"
bothers me. Surely, the $C_{60}$ is hardly preferred from the other $C_n$
molecules in the laser pulse then grows under the non-violent
(cool) conditions of the expanding beam? Of course, this does
not preclude the possibility that the $C_{60}$ could grow in the cool,
dense atmospheres of carbon-rich stars. Finally, dare the
authors speculate as to the likely form of the IR spectrum of the
$C_{60}$ molecule (compared to say graphite)?! The Letter should be
published in Nature since I feel that the subject matter will be
of interest to people from several disciplines.

**Figure 2** Comments of the referees on the paper accepted by *Nature* on
18 October 1985.

road, travelled in a parallel abstract universe by a small,
disparate band of imaginative theoreticians, all seemingly
unaware of each others' efforts. I went to a spectroscopy
conference in Riccione, Italy, scarcely a week after the
discovery and excitedly announced, in a very short special

three-minute presentation, that we had observed a strong signal for a species with 60 carbon atoms and had proposed the truncated icosahedral closed cage structure for the species. A former student of mine, Julie August, who was present at the conference returned to Nottingham University with the news and her then supervisor, Martyn Polliakoff, wrote to tell me that his friend David Jones had, in a highly imaginative 1966 *New Scientist* article under the pseudonym Daedalus,[59*,60] already described how an extended graphene sheet consisting of hexagonal network would close if 12 pentagonal disclinations can somehow be introduced among the hexagons intrinsic to graphite. For me there was a crucial revelation in David's article; this was the fact, which is a consequence of Euler's law, that the insertion of 12 pentagons among any number of hexagons is necessary and sufficient to close a network. I immediately realised that a minimum of 60 carbon atoms was needed to close a cage if abutting pentagons are to be avoided. This requirement together with the fact that pentalene $C_8H_6$ (which consists of two fused pentagons) is unstable, indeed antiaromatic as generally discussed in undergraduate Quantum Chemistry courses, was the first piece of circumstantial evidence supporting our structural proposal. It was for me the sort of convincing evidence I needed to convince myself that our structure proposal must be correct, and I would not have to commit suicide! David has also kindly penned a few notes describing how he came upon this highly original conjecture **(A6)**. Within a few more days we discovered that David's paper was not the only one discussing pure carbon cages. My friend and colleague Nenad Trnajstic at the Institut Rudjer Boskovic in Zagreb informed me that there were some more papers: Eiji Osawa had in 1970 published the idea of $C_{60}$ itself[61] in Japanese. The following year the idea was further elaborated on in his book *Aromaticity*, written with Zenichi Yoshida.[62] This appears to be the very first publication ever on $C_{60}$. Eiji has kindly also contributed a personal account detailing how he came to his highly imaginative conclusion that $C_{60}$ might be stable and possibly aromatic (**A7a**) and has translated the section of his book[62] dealing with $C_{60}$ (**A7b**). Nenad also informed me that in 1972, two years after Eiji's publication, Bochvar and Gal'pern had published the Hückel calculation for $C_{60}$.[63] Then we discovered that Davidson in 1980[64] had also published a Hückel analysis. Just after news of our discovery became known, Fritz Shaeffer at Berkeley informed us that Tony Haymet had just a few months previously, given a presentation on $C_{60}$. Haymet published his study a few months later.[65] As well as this set of highly imaginative quantum chemists who had been ruminating over the $C_{60}$ molecule, there was at least one experimental synthetic research project, that by Orville Chapman at UCLA, which had focused on devising a traditional synthetic strategy to create $C_{60}$.[66] This daunting task was finally successfully accomplished by Larry Scott et al. in 2002.[67] Larry has also written a fascinating account of his journey which gives a valuable insight into the way this beautiful molecule can stimulate new science (**A8**). Sean in his personal account (**A5**) describes how the Rice group was apprised of some previous work and argues, with some reason, that we should have been more careful before submitting the manuscript for publication and made some effort to check whether there was any previous literature on $C_{60}$! Of course, that is far easier today with the Internet than it was in 1985.

Thus in retrospect, in chronological order, the key advances and/or observations which underpinned the basic discovery experiment were (a) the detection of the mass spectra of large carbon species with up to 33 atoms in the product of a carbon arc discharge by Hintenberger and coworkers, (b) the synthesis and microwave study of cyanopolyynes at Sussex and their subsequent detection in the interstellar medium by the Sussex/NRC Canada radioastronomy collaboration, (c) the development of the laser vaporisation supersonic cluster beam mass spectrometric technique by Smalley's Rice University group which revolutionised the study of refractory clusters, and (d) the discovery of the carbon star IRC+10216 by Becklin, Neugebauer and colleagues and the subsequent discovery that our cyanopolyynes were being ejected from IRC+10216 into the interstellar medium by Winnewisser and Walmsley.

## 7. The Circumstantial Supporting Evidence

After the discovery, numerous projects were devised to prove, either by extraction or otherwise, our contention that $C_{60}$ was indeed the buckminsterfullerene structure we had proposed. Joint Rice/Sussex studies were carried out as well as studies by the Rice and Sussex groups independently, and in time an overwhelming body of circumstantial evidence was assembled. Jim Heath was particularly heavily involved in the follow-up work at Rice as well as the first experiments. Some of the key observations, made during the period after discovery and prior to the Krätschmer et al. paper[2] in 1990, which gave the strongest supporting evidence for the buckminsterfullerene structure were:

(i) The creation of the lanthanum complex La@$C_{60}$ from which it was shown that it was impossible to dislodge the La atom, so providing good evidence that the La atom was not on the outside but possibly trapped inside the putative cage.[68*] In addition, in this paper a rather neat structure for $C_{70}$ was proposed in which a ring of 10 carbon atoms was inserted between two $C_{30}$ half-domes, fitting in nicely with the fact that $C_{70}$ was clearly the second favoured entity experimentally.

(ii) The Isolated Pentagon Rule (IPR) to rationalise the strengths of $C_{60}$ and $C_{70}$ as well as the Isolated Multiplet Pentagon Rule (IMPR) to rationalise the prominence of other features such as $C_{28}$ and $C_{50}$ were proposed.[69*] These rules had been derived on the basis of simple molecular-model construction together with knowledge of aromatic and anti-aromatic reactivity. Tom Schmalz, William Seitz, Doug Klein and Gerald Hite in their very nice simultaneous much more detailed theoretical study[70*] proved the IPR and IMPR conjectures theoretically. In fact this advance was, at least for me, absolutely conclusive, as it indicated that *if $C_{60}$ were actually a cage, then $C_{70}$ had to be the second favoured species because no IPR structure existed between $C_{60}$ and $C_{70}$.* I remembered thinking that although another solution might exist for a 60-atom species, there was no way that another solution would cough up both $C_{60}$ and $C_{70}$.

(iii) An important result of the generalised IMPR rule[69*,70*] was the proposal that a $C_{28}$ species should be a tetravalent "superatom" with a tetrahedral structure (**Fig. 3a**) which has special stability.[69*] This result was corroborated by mass spectroscopic data such as that shown in **Fig. 3b**, where $C_{28}$ can be shown under certain clustering conditions to be special.

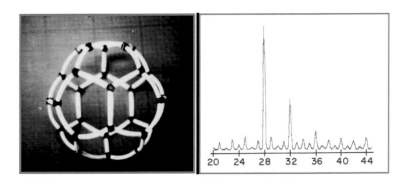

**Figure 3** (a) Polaroid image of the first molecular model of $C_{28}$ photographed where it was constructed, on our coffee table[69*] at home in Sussex. (b) Carbon cluster mass spectrum under certain conditions in the $C_{20}$–$C_{44}$ range (Rice/Sussex unpublished data 1985).

(iv) The so-called "Shrink-Rap" experiment[71] showed that on laser photodissociation M@$C_{60}$ lost $C_2$ fragments sequentially until the M@C$n$ species of a given minimum $n$ fragmented completely into small carbon species when the cage became too small to encase the enclosed metal atom completely. The size of the endohedral complex prior to complete fragmentation increased commensurate with the known size of the encapsulated atom.

(v) In the first study of giant fullerenes it was deduced on the basis of simple model building that they exhibited quasi-icosahedral shapes **(Fig. 4a,b)** and this result explained the structures of previously puzzling electron microscope images of certain macroscopic carbon particles as consisting of concentric fullerene cages in onion-like or Russian Doll configurations.[72*,73]

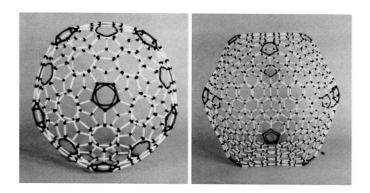

**Figure 4** (a, b) Photographs of the beautiful molecular models $C_{240}$ and $C_{540}$, in the first study of giant fullerenes; constructions by Ken McKay at Sussex.[72*] Note how the curvature is focused in the regions of the pentagonal (red) disclinations and the overall shape becomes more closely icosahedral as the number of atoms in the cage increases.

(vi) In an important paper[74] it was shown that the $C_{60}$ species we had detected was special whether positively or negatively charged or indeed neutral, contrary to one claim.[75]

During this period I developed my **"4/5 Rule of Scientific Reliability"**:

*If one has a hypothesis one should carry out at least five different studies to ascertain whether the hypothesis is correct or not: If 4 out of 5 studies fit the hypothesis, it is* almost *certainly correct (with the emphasis on the word* almost*); if on the other hand only 1 out of 5 fits, the hypothesis is* almost *certainly wrong (again with the accent on the word* almost!*). As a corollary, if only 1 out of 5 did not fit, it will usually be found that that study had been subject to some error and subsequently found to fit under more careful scrutiny.*

## 8. Skeptical Studies

It is probably hard for young chemistry students today, for whom the buckminsterfullerene has ever been an elegant iconic star, embedded in the rich tapestry of the Chemical Universe rather like Halley's comet in the famous Bayeux Tapestry or the painting by Giotto, to realise that for some the buckminsterfullerene structure was, at the time it was proposed, somewhat bewildering and highly unconvincing. The fact that it happened at roughly the same time as the "Cold Fusion" saga did not help either. Indeed the proposal appeared to be so revolutionary that several papers were published which were strongly critical not only of our structural conjecture but also our experimental observations. A footnote in the first skeptical paper titled "$C_{60}La$: a Deflated Soccerball?"[76] indicated that a detailed study, which was to be published later, would show that the strength of the $C_{60}$ signal which we had detected was not an indication of special character but rather an artefact which depended on a complex "interplay" of various experimental parameters involved in our measurement. These claims were totally at variance with the observations which we had by then made and when the paper, mentioned in the footnote, was subsequently published,[77] it did not support the negative claims made. A claim, made in another skeptical paper, that the negative carbon ion distribution did not support the special nature of $C_{60}$[75] was refuted unequivocally by a careful exhaustive study which we carried out in response.[74] Paradoxically, a mechanistic transformation published in skeptical theoretical paper[78] which poured even more doubt on our buckminsterfullerene structural proposal is now used frequently to explain mechanistic aspects of fullerene dynamics! A few more skeptical papers were published but by and large, as time passed, the onslaught was repulsed and gradually the increasing weight of supportive studies by us, together with those of numerous other groups such as the fragmentation study by Eleanor Campbell, Ingolf Hertel and coworkers,[79] indicated overwhelmingly that $C_{60}$ was indeed a stable entity and that the structure was almost certainly a truncated icosahedron. The set of skeptical papers represents yet another typical example of the fact that some breakthroughs, which appear to conflict with received wisdom, can encounter significant negative criticism before becoming accepted. They also serve as an object lesson on how extremely careful one should be when criticising the work of others!

## 9. Spectroscopic Theory

Many researchers, in particular theoreticians, embraced the new proposal enthusiastically and were stimulated to examine some important likely properties of $C_{60}$. Of the many early studies discussed in the review paper,[4] two papers which appear to have been influential in the extraction breakthrough by the Heidelberg/Arizona team serve to summarise two key theoretical results:

(a) Theoretical studies such as that by Weeks and Harter[80] indicated that only 4 vibrational modes should be infrared active and provided approximate and fairly accurate associated infrared frequencies.

(b) Some studies such as that of Larsson et al.[81] predicted roughly that the UV spectrum of $C_{60}$ should lie fairly close to the 217 nm feature described above.

### 10. Extraction and Structure Proof

As indicated above, theoreticians found the fullerene conjecture fascinating and produced the key predictions (a) and (b) in the above paragraph on the vibrational and electronic spectra which proved to be important in the work of Wolfgang Krätschmer, Lowell Lamb, Kostas Fostiropoulos

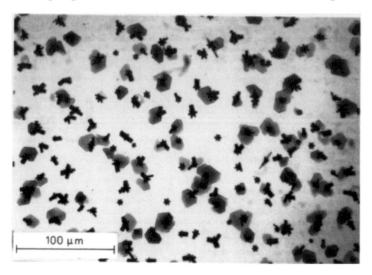

100 μm

**Figure 5** The amazing (at the time) photograph of crystalline carbon produced by Krätschmer and coworkers.[2] Photograph courtesy of Wolfgang Krätschmer. One of the most important images in the whole of science.

and Donald Huffman, who announced in 1990 that they had extracted macroscopic amounts of $C_{60}$.[2] I do not think there is any more elegant story in the whole of chemistry in the latter half of the 20th century than the detective story which led on from the early Day/Huffman 1977 study of carbon dust[48] via the proposed infrared spectrum of $C_{60}$ to success in extraction. I also think that this must be the most historically important example of the use of infrared finger-printing in an important chemical advance. This experiment vanquished all criticism by producing as evidence images of crystals extracted from the deposit of arc-discharged carbon which X-ray analysis indicated consisted of arrays spheroidal carbon molecules almost exactly as expected, 1 nm apart, centre to centre.

Simultaneously, in an independent study the 720 amu mass spectrum of $C_{60}$ was measured at the University of Sussex by Ala'a Abdul Sada from a deposit of arc-discharged carbon. A red solution was extracted a few weeks later by Jonathan Hare and five days prior to the submission of the Krätschmer et al. paper[2] for publication! The one-line NMR spectrum was also measured by Tony Avent, thus proving conclusively, once and for all, that all the carbon atoms were identical and the truncated icosahedral structure was the only possible solution.[3]

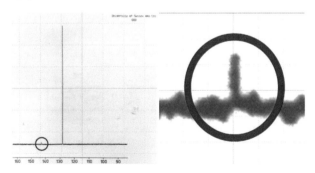

**Figure 6** The first detection of the NMR line of $C_{60}$ is the tiny blip at ca $\delta = 143$. Reproduced from Ref. 3 with permission from The Royal Society of Chemistry. Small is, as they say, beautiful!

The chromatographic method of separating various members of the fullerene family was developed at Sussex by my late colleague Roger Taylor with advice from another our colleagues, Jim Hanson. The final absolute and unequivocally definitive proof was provided by the detection of a five-line NMR spectrum of the chromatographically isolated sample of $C_{70}$.[3]

**Figure 7** The first pure samples of $C_{60}$ and $C_{70}$ in benzene solution separated by chromatography at the University of Sussex. Reproduced from Ref. 3 with permission from The Royal Society of Chemistry. In benzene solution $C_{60}$ is magenta and $C_{70}$ on the right red.

In the case of the Sussex group an earlier electron microscope study of arc-discharged carbon had indicated that the morphology of the resulting carbon deposit changed dramatically from an even, flat, very thin layer deposit to a floccular one as the pressure of argon gas under which it was formed increased. Further work on this project, which had suggested perhaps that roundish carbon structures like $C_{60}$ might be involved, was thwarted by the lack of financial support. It was resurrected when Michael Jura, an astrophysicist friend and colleague at UCLA, sent me some very interesting preliminary results that Krätschmer et al. had reported at an astrophysical conference in Capri.[82] As recounted above, a red solution of a $C_{60}/C_{70}$ extract was produced from arc-discharged graphite rods five days before the Krätschmer et al. manuscript arrived at Sussex from *Nature* for review detailing their brilliant results.[2] Lowell Lamb and Jonathan Hare were research students in the Krätschmer et al. group and our Sussex group respectively and their personal accounts, **(A9)** and **(A10)** respectively, published here give a delightfully intimate insight into the excitement that they experienced as young researchers during those heady days.

In retrospect we can see that the original work on carbon smoke by Day and Huffman[48] and the later detection of the UV and infrared spectra[2,82,A10] together with the post-discovery theoretical work on the electronic and vibrational spectra of $C_{60}$[80,81] were crucial factors in the brilliant detective story surrounding the extraction of $C_{60}$ by the Heidelberg/Arizona group.

## 11. Aftermath Studies

At least one organic chemist, Orville Chapman at UCLA, had initiated a serious effort to create $C_{60}$ by a traditional synthetic route.[66] This non-trivial feat was finally accomplished by Larry Scott and his group.[67] First Larry developed an elegant alternative route to the bowl-shaped molecule corannulene $C_{20}H_{12}$, which is a pentagonal ring surrounded by five hexagonal rings.[83] Larry's new strategy was simpler than that in the seminal early work of Barth and Lawton.[84] In doing so Larry has started to map out the strategy for the engineering of extended carbon nanostructure which I think will be necessary if carbon-based nanotechnology is to fulfil the fantastic promise that electrical and tensile strength studies imply are possible. Such advances are only going to be achieved if absolutely perfect carbon nanostructures can be realised on a consistent bulk scale—for instance, bundles of perfect hexagonally packed single-walled nanotubes all of the same diameter and preferably same chirality. Larry has also kindly agreed to pen an article which highlights the thinking and strategy development behind this beautiful advance (**A8**).

The result of all the above studies has been an explosion of research into the chemistry and physics of the fullerenes which now runs at approximately a thousand papers per year. Further study of the material produced by the Krätschmer/Huffman technique resulted in the discovery by Sumio Iijima[85] that nanotubes could also be produced and the realisation that they had in fact been observed some years before by Morinobu Endo and co-workers.[86] In fact the discovery of $C_{60}$ re-awakened intense interest in fundamental aspects of carbon chemistry and carbon materials science. In particular it ignited the field of carbon nanoscience and nanotechnology and may even have created the new tsunami of interest in carbon science which led to the most recent explosion of research into graphene.[87]

A paper in 1993 on $U@C_{28}$[88] presented interesting support for the prediction I made five years earlier in 1987,[69] that the small fullerene $C_{28}$ (**Fig. 3**) should be (a) special, (b) tetrahedral and (c) tetravalent. Recent highly detailed experiments utilising the remarkable resolution of the FT-ICR-MS developed by Alan Marshall at FSU have shown that the species $M@C_{28}$ (M = Ti, Zr and U) are indeed very special,[89] providing unequivocal empirical evidence for the predicted special nature of $C_{28}$. Paul Birkett at Sussex, working with my colleagues Roger Taylor and David Walton, discovered that $C_{60}$ could add six Cl atoms in a fascinating symmetric pattern.[90] This result can be considered to be basically a fullerene addition reactivity pattern analogous to the ortho-, meta- and para- directivity observed in the reactivity of benzene upon substitution. Robert Haddon, then at Bell labs, and colleagues discovered that K intercalated in crystalline $C_{60}$ was superconducting at the relatively high temperature of 18 K[91] and this led to a flurry of interest in organic superconductivity. Work at UCLA by Fred Wudl's group led to the discovery of the fact that nitrogen could replace a carbon atom to form a $C_{59}N$ analogue which forms the stable dimer $(C_{59}N)_2$.[92] At FSU we have used the FT-ICR-MS to study the way that B can displace a C atom in $C_{60}$ to form the species $C_{59}B$[93] which has so far not been isolated. Perhaps one of the most interesting results was obtained by Mike Bowers and his coworkers, who showed, using an ion drift tube technique, that in the range $C_n$ ($n$ = 1–60) there are at least five different families of pure carbon species[94] in carbon vapour! One family is the linear chains and another set monocyclic rings which have been reliably identified by John Maier's group in Basel.[95] We of course identified the third the closed-cage fullerenes but two other families in the drift tube study still have to be identified with a significant degree of certainty.

## 12. Mechanism

As far as post-1990 fullerene research is concerned, it is hard to pick out all the key breakthroughs among some 20,000 or more papers but from a personal perspective there are a few. There have been numerous papers ruminating over various possible mechanisms for the formation of fullerenes and their elongated analogues the nanotubes, and the situation is certainly complicated. Two papers which suggested that closed-cage growth might be involved in the case of $C_{60}$ and the nanotubes were published in 1992 by Jim Heath[96] and by Morinobu Endo and me,[97] respectively. A closed-cage mechanism has now been unequivocally confirmed experimentally[98] as operating under the same conditions as those under which $C_{60}$ was discovered in the original experiment.[1] The experiments were carried out at the National High Field Magnet Laboratory (NHFML) at Florida State University (FSU) using the exquisite state-of-the-art resolution afforded by the Fourier transform ion cyclotron resonance mass spectrometer (FT-ICR-MS) developed by Alan Marshall. These experiments showed that carbon cages can indeed "ingest" small carbon species such as C, $C_2$, etc., to form larger and larger fullerenes.[98] Paul Dunk has written his personal account of cluster experiments[89,98] showing again how important youthful exuberance and imagination are to the propagation of fundamental research (**A11**). A study published in 1992 by Jack Howard's group indicated that $C_{60}$ can be readily extracted from hydrocarbon flames.[99] It might be noted that in 1986, some six years earlier, we published a paper[100] in which it was suggested that $C_{60}$ might be involved in some way in combustion processes during the soot formation phase. This conjecture was severely criticised at the time by Frenklach and Ebert.[101] In this context it may be worthy of some careful thought that (a) hydrocarbon flames have been studied assiduously since at least the time of Michael Faraday, and as far as we know, no single combustion scientist either discovered fullerenes in their flames or predicted their presence theoretically prior to our discovery,[1] and (b) the Mitsubishi Chemical Company extracts $C_{60}$ in kilogram quantities from the soot produced by the combustion of (I understand) methane! One obvious conclusion one might draw from these observations is that, even after all the decades of research into soot formation, the hydrocarbon combustion process which creates the soot from which fullerenes are extracted in commercial quantities is not understood at all and needs to be explained!

### 13. Epilogue: Detection of $C_{60}$ in Space

Finally, as described above, the quest for an understanding of the existence of our long carbon chain molecules in space was the catalyst of the experiment using Rick Smalley's cluster beam apparatus which revealed that $C_{60}$ self-assembled spontaneously under conditions which simulated the conditions in red giant carbon stars such as IRC+10216. This led, at least in my mind, to the conviction that $C_{60}$ must be lurking in space.[23,40,41] Furthermore, the probability that the molecule existed in space and the possibility that it might be involved in some way with the diffuse interstellar bands were proposed in a paper with Michael Jura.[102] In this paper we suggested that some $C_{60}$ molecules in space were highly likely to have attached atoms and if these atoms were relatively abundant such as Na, Mg, Ca, etc., this should give rise to relatively strong charge-transfer bands which might account for some of the DIBs. In 2010 the conjecture that $C_{60}$ must be in space was beautifully, indeed amazingly, confirmed by Jan Cami, Jeronimo Bernard-Salas, Els Peeters and Sarah Elizabeth Malek,[103] who were able to assign all the detected features in a Spitzer Telescope IR spectrum of a young planetary nebula to $C_{60}$ and $C_{70}$. Thus the story has turned full circle twice: from the synthesis in the laboratory of carbon molecules ca. 1 nm in length to the detection of these molecules in various regions of the interstellar medium and then back to a terrestrial laboratory to uncover the existence of a 1 nm diameter soccerball-shaped all-carbon object and now back into space yet again. It has been a tremendously exciting roller coaster research ride for almost everyone involved, and what is more, the ride is not yet over and if anyone thinks it might be, a casual skim through the pages of *An Atlas of Fullerenes*, by Patrick Fowler and David Manalopoulos,[104] might just change their mind!

## References

*Note:* References in grey are reproduced in Part C.

1. H W Kroto, J R Heath, S C O'Brien, R F Curl and R E Smalley, *Nature*, **318**, 162–163 (1985).
2. W Krätschmer, L Lamb, K Fostiropoulos and D R Huffman, *Nature*, **347**, 354–358 (1990).

3. R Taylor, J P Hare, A K Abdul-Sada and H W Kroto, *J Chem Soc Chem, Commun*, 1423–1425 (1990).

4. H W Kroto, A W Allaf and S P Balm, *Chem Rev*, **91**, 1213–1235 (1991).

5. R F Curl and R E Smalley, *Science*, **242**, 1017–1022 (1988).

6. H W Kroto, *Science*, **242**, 1139–1145 (1988).

7. R F Curl and R E Smalley, *Scientific American*, 32 (Oct 1991).

8. H W Kroto, *Angew Chem Int Edit*, **31**, 111–129 (1992).

9. R F Curl, *Rev Mod Phys*, **69**, 691–702 (1997), from Nobel Lecture, 187–231 (1996).

10. H W Kroto, *Rev Mod Phys*, **69**, 703–722 (1997), from *Les Prix Nobel 1997*, Nobel Lecture, 187–231 (1996).

11. R E Smalley, *Rev Mod Phys*, **69**, 723–730 (1997), from *Les Prix Nobel 1997*, Nobel Lecture, 187–231 (1996).

12. W Krätschmer, in *The Physics of Fullerene-Based and Fullerene-Related Materials*, ed. W Andreoni (2000), Kluwer.

13. E E B Campbell and I V Hertel, "The physics of fullerenes in the gas phase," in *Von Fuller bis zu Fullerenen*, ed. W. Krätschmer und H. Schuster Vieweg Verl., Heidelberg, 143–160 (1996).

14. W Krätschmer, in *The Building Blocks of Next Generation Nanodevices*, ed. V K Prashant, D M Guldi and F. D'Śouza, The Electrochemical Society (2003).

15. K Fostiropoulos, *Int J Mod Phys B*, **06**, 3791–3800 (1992).

16. D R Huffman, *Physics Today*, **44**, 22–29 (1991).

17. G Taubes, *Science*, **253**, 1476–1479 (1991).

18. J Baggott, *Perfect Symmetry: Accidental Discovery of Buckminsterfullerene*, Oxford University Press (1994).

19. H Aldersey Williams, *The Most Beautiful Molecule: The Discovery of the Buckyball*, Wiley (1997).

20. I Hargittai, *Candid Science: Conversations with Famous Chemists*, Imperial College Press (2000).

21. E J Applewhite, *The Chemical Intelligence*, (July 1995), Vol 1 No. 3, Springer, http://www.4dsolutions.net/synergetica/eja1.html

22. A Nickon and E F Silversmith Organic Chemistry, *The Name Game: Modern Coined Terms and Their Origins*, Pergamon, New York (1987).

23. BBC Horizon UK TV film *Molecules with Sunglasses* (also Nova US) (1993), http://www.youtube.com/watch?v=xVZRGcg-BXI

24. H Hintenberger, J Franzen and K D Schuy, *Z Naturforsch*, **18a**, 1236–1237 (1963).

25. R Campargue, *J Chem Phys*, **52**, 1795–1802 (1970).

26. R Campargue, 24th International Symposium on Rarefied Gas Dynamics, AIP Conf. Proc. 762, 32–46 (2004).

27. R E Smalley, L Wharton and D H Levy, *J Chem Phys*, **63**, 4977–4989 (1975).

28. T G Dietz, M A Duncan, D E Powers and R E Smalley, *J Chem Phys*, **74**, 6511–6512 (1981).

29. L Michalopoulos, M E Geusic, P R R Langridge-Smith and R E Smalley, *J Chem Phys*, **80**, 3556–3560 (1984).

30. R Eastmond and D R M Walton, *Chem Comm*, 204–205 (1968).

31. R Eastmond and D R M Walton, *Tetrahedron*, **28**, 4601–4616 (1972).

32. T R Johnson and D R M Walton, *Tetrahedron*, **28**, 5221–5236 (1972).

33. A J Alexander, H W Kroto and D R M Walton, *J Mol Spectrosc*, **62**, 175–180 (1976).

34. L W Avery, N W Broten, J M MacLeod, T Oka and H W Kroto, *Astrophys J*, **205**, L173–L175 (1976).

35. C Kirby, H W Kroto and D R M Walton, *J Mol Spectrosc*, **83**, 261 (1980).

36. H W Kroto, C Kirby, D R M Walton, L W Avery, N W Broten, J M MacLeod and T Oka, *Astrophysics J*, **219**, L133–L137 (1978).

37. N W Broten, T Oka, L W Avery, J M MacLeod and H W Kroto, *Astrophys J*, **223**, L105–L107 (1978).

38. E E Becklin, J A Frogel, A R Hyland, J Kristian, and G Neugebauer, *Astrophys J*, **158**, L133–L137 (1969).

39. G. Winnewisser and C. M. Walmsley, *Astrophys J*, **70**, L37–L39 (1978).

40. H W Kroto, *Int Rev Phys Chem*, **1**, 309–376 (1981).

41. H W Kroto, *Chem Soc Rev*, **11**, 435–491 (1982).

42. M L Heger, *Lick Observatory Bulletin*, **337**, 141–145 (1922).

43. A McKellar, *Pub Dom Ap Obs*, **7**, 251 (1942).

44. G H Herbig, *Astrophys J*, **196**, 129–160 (1975).

45. G H Herbig, *Annu Rev Astron Astrophys*, **33**, 19–74 (1995).

46. A E Douglas, *Nature*, **269**, 130–132 (1977).

47. T P Stecher, *Astrophys J*, **157**, L125 (1969).

48. K L Day and D R Huffman, *Nature*, **243**, 50 (1973).

49. P M Solomon and W Klemperer, *Astrophys J*, **178**, 389 (1972).

50. E Herbst and W Klemperer, *Astrophys J*, **185**, 505–533 (1973).

51. J M Greenberg, *Surf Sci*, **500**, 793–822, (2002).

52. E A Rohlfing, D M Cox and A Kaldor, *J Chem Phys*, **81**, 3322–3330 (1984).

53. A G Whittaker, *Science*, **200**, 763 (1978).

54. H W Kroto, *Chemistry World* (Nov 2010), http://www.rsc.org/chemistryworld/Issues/2010/November/CarbyneOtherMythsAboutCarbon.asp

55. P P K Smith and P R Buseck, *Science*, **216**, 984–986 (1982).

56. L A Bloomfield, M E Geusic, R R Freeman, W Brown, *Chem Phys Lett*, **121**, 33–37 (1985).

57. J R Heath, Q Zhang, S C O'Brien, R F Curl, H W Kroto and R E Smalley, *J Am Chem Soc*, **109**, 359–363 (1987).

58. H W Kroto, J R Heath, S C O'Brien, R F Curl and R E Smalley, *Astrophys J*, **314**, 352–355 (1987).

59. D E H Jones, *New Scientist*, **245**, 118–119 (1966) [Daedalus column].

60. D E H Jones, *The Inventions of Daedalus*, Freeman, Oxford (1982).

61. E Osawa, *Kagaku*, **25**, 854–863 (1970) [in Japanese].

62. Z Yoshida and E Osawa, *Aromaticity*, Kagakudojin (1971) [in Japanese].

63. D A Bochvar and E G Gal'pern, *Dokl Akad Nauk USSR*, **209**, 610–612 (1973); *Proc Acad Sci USSR*, **209**, 239–241 (1973) [in English].

64. R A Davidson, *Theor Chim Acta*, **58**, 193–195 (1981).

65. A D J Haymet, *Chem Phys Lett*, **122**, 421–424 (1985).

66. O Chapman [cf. ref. 18 book by Baggott and ref. 23 BBC Horizon film].

67. L T Scott, M M Boorum, B J McMahon, S Hagen, J Mack, J Blank, H Wegner, S de Meijere, *Science*, **295**, 1500–1503 (2002).

68. J R Heath, S C O'Brien, Q Zhang, Y Liu, R F Curl, H W Kroto, F K Tittel and R E Smalley, *J Am Chem Soc*, **107**, 7779–7780 (1985).

69. H W Kroto, *Nature*, **329**, 529–531 (1987).

70. T G Schmalz, W A Seitz, D J Klein and G E Hite, *J Am Chem Soc*, **110**, 1113–1127 (1988).

71. F D Weiss, J L Elkind, S C O'Brien, R F Curl and R E Smalley, *J Am Chem Soc*, **110**, 4464–4465 (1988).

72. H W Kroto and K G McKay, *Nature*, **331**, 328–331 (1988).

73. K G McKay, H W Kroto and D J Wales, *J Chem Soc, Faraday Trans*, **88**, 2815–2821 (1992).

74. S C O'Brien, J R Heath, H W Kroto, R F Curl and R E Smalley, *Chem Phys Lett* **132**, 99–102 (1986).

75. M Y Hahn, E C Honea, A J Paguia, K E Schriver, A M Camarena and R L Whetten, *Chem Phys Lett*, **130**, 12–16 (1986).

76. D M Cox, D J Trevor, K C Reichmann and A Kaldor, *J Am Chem Soc*, **108**, 2457–2458 (1986).

77. D M Cox, K C Reichmann and A Kaldor, *J Chem Phys*, **88**, 1588–1597 (1988).

78. A J Stone and D J Wales, *Chem Phys Lett*, **128**, 501–503 (1986).

79. E E B Campbell, G Ulmer, H–G Busmann and I V Hertel, *Chem Phys Lett*, **175**, 505–510 (1990).

80. D E Weeks and W G Harter, *J Chem Phys*, **90**, 4744–4771 (1989).

81. S Larsson, A Volosov and A Rosen, *Chem Phys Lett*, **137**, 501 (1987).

82. W Krätschmer, K Fostiropoulos and D R Huffman, in *Dusty Objects in the Universe*, ed. E Bussoletti and A A Vittone, Kluwer, 89–93 (1989).

83. L T Scott, M M Hashemi, D T Meyer and H B Warren, *J Am Chem Soc*, **113**, 7082–7084 (1991).

84. W E Barth and R G Lawton, *J Am Chem Soc*, **88**, 380–381 (1966).

85. S Iijima, *J Cryst Growth*, **50**, 675–683 (1980).

86. A Oberlin, M Endo and T Koyama, *J Cryst Growth*, **32**, 335–349 (1976).

87. A K Geim and K S Novoselov, *Nat Mater*, **6**, 183–191 (2007).

88. T Guo, M D Diener, Y Chai, M J Alford, R E Haufler, S M Mcclure, T R Ohno, J H Weaver, G E Scuseria and R E Smalley, *Science*, **257**, 1661–1664 (1992).

89. P W Dunk, N K Kaiser, M Mulet–Gas, A Rodríguez–Fortea, J M Poblet, H Shinohara, C L Hendrickson, A G Marshall, and H W Kroto, *J Am Chem Soc*, **134**, 9380–9389 (2012).

90. P R Birkett, A G Avent, A D Darwish, H W Kroto, R Taylor and D R M Walton, *J Chem Soc, Chem Commun*, 1230 (1993).

91. R C Haddon, A F Hebard, M J Rosseinsky, D W Murphy, S J Duclos, K B Lyons, B Miller, J M Rosamilia, R M Fleming, A Kortan, S H Glarum, A V Makhija, A J Muller, R H Eick, S M Zahurak, R Tycko, G Dabbagh, F A Thiel, *Nature*, **350**, 320–322 (1991).

92. J C Hummelen, B Knight, J Pavlovich, Ro González and F Wudl, *Science*, **269**, 1554–1556 (1995).

93. P W Dunk, A Rodríguez-Fortea, N K Kaiser, H Shinohara, J M Poblet, and H W Kroto, *Angew Chem Int Ed*, **52**, 315–319 (2013).

94. G von Helden, M-T Hsu, P R Kemper and M T Bowers, *J Chem Phys*, **95**, 3835–3837 (1991).

95. H Ding and J P Maier, *J Phys Conf Ser*, **61**, 252 (2007).

96. J R Heath, in *Fullerenes: Synthesis, Properties and Chemistry of Large Carbon Clusters*, ACS Symposium Series No. 481, ed. G S Hammond and V J Kuck, 1–23 (1992).

97. M Endo and H W Kroto, *J Phys Chem*, **96**, 6941–6944 (1992).

98. P W Dunk, N K Kaiser, C L Hendrickson, J P Quinn, C P Ewels, Y Nakanishi, Y Sasaki, H Shinohara, A G Marshall and H W Kroto, *Nat Commun*, **3**, 855 (2012).

99. J B Howard, J T McKinnon, Y Makarovsky, A L Lafleur, and M E Johnson, *Nature*, **352**, 6331 (1992).

100. Q L Zhang, S C O'Brien, J R Heath, Y Liu, R F Curl, H W Kroto and R E Smalley, *J Phys Chem*, **90**, 525–528 (1986).

101. M Frenklach and L B Ebert, *J Phys Chem*, **92**, 561–563 (1988).

102. H W Kroto and M Jura, *Astron Astrophys*, **263**, 275–280 (1992).

103. J Cami, J Bernard-Salas, E Peeters and S E Malek, *Science*, **329**, 1180 (2010).

104. P W Fowler and D E Manalopoulos, *An Atlas of Fullerenes*, Oxford (1995).

# Part B

# History of "the nozzle"

## Lennard Wharton

*Evidentia Engineering Inc., 10 Park Pl, Short Hills, NJ 07078, USA*
len.wharton@verizon.net

My research at Harvard and later at the University of Chicago obtained high-resolution radio frequency and microwave spectra of simple molecules, using molecular beam electric resonance equipment that we designed and built ourselves. The technique selected molecules in a few low rotational states from a molecular beam using electric quadrupole state selectors, in equipment adapted from designs by John Triska at Syracuse and Christophe Schlier at Freiburg. The idea to use molecular beams to study electrical properties of molecules of chemical interest came from my research sponsor at Harvard, William Klemperer, who was a young assistant professor without tenure at the time. We were fortunate to have enthusiastic support from Norman Ramsey across Oxford Street in the physics department.

*C60 Buckminsterfullerene: Some Inside Stories*
Edited by Harry Kroto
Copyright © 2015 Pan Stanford Publishing Pte. Ltd.
ISBN 978-981-4463-71-3 (Hardcover), 978-981-4463-72-0 (eBook)
www.panstanford.com

At Harvard we set up our stainless steel molecular beam equipment in a former darkroom in the basement of Mallinckrodt Laboratories next to George Kistiakowsky's glass mass spectrometer. Kisty would come by every other weekend or so from Washington (where he was science advisor to President Eisenhower) and on occasion look at our metal apparatus being slowly built, shake his head and say in good humor in his Russia-accented English, "Molecular beams are the death of the chemist." A paper in the *Review of Scientific Instruments* "reported a not very successful attempt by Kistiakowsky and Schlichter to reduce the Kantrowitz-Grey idea [skimmed supersonic jet] to practice."[1] Ironically, Kisty's paper was published around the same time that other investigators were beginning to reduce high-intensity molecular beam sources to practice.[2]

But Kisty played a role in the post-Sputnik flood of new federal funding in the physical sciences, and for that I am grateful. It helped to pay for the development of our molecular beam.

At Chicago my research group of students and postdocs attempted (unsuccessfully) to use electric dipoles to create a molecular accelerator for studying gas–gas and gas–surface scattering collisions at elevated translational energy. Having failed at acceleration, I turned our equipment to investigate such collisions at non-accelerated—thermal—energies. In the collision scattering apparatus, the primary molecular beam was a highly velocity selected effusion beam instead of an accelerator and the target was a perpendicular supersonic jet.

In 1968–1970, we developed a high-intensity supersonic target jet that could handle "non-condensable" scattering gases. This target enabled us to observe at high angular

---

[1]John Fenn, *Autobiography*. 2002 Nobel Prize.
[2]John Fenn, op cit.

and velocity resolution the differential gas–gas scattering patterns that showed beautiful de Broglie wave diffraction and rainbow effects. The "non-condensable" target gases included argon, nitrogen, and carbon monoxide. I put "non-condensable" in quotation marks because the target equipment in fact condensed these gases, i.e. froze them out, on copper surfaces maintained at 20 K cooled with liquid hydrogen.

The supersonic jet scattering gas target was dense and had a unique velocity. The use of cryo-pumped surfaces was to enable it to be mechanically compact and self-contained with its own high-speed pumping system in order to fit it into a scattering apparatus. To be capable of pumping the large quantities of a "non-condensable" gas flowing from a high-pressure (5 bar) supersonic jet into a high-vacuum scattering chamber, liquid hydrogen cryo-pumping was used for the three chambers in the target: between the nozzle and the skimmer, between the skimmer and a collimator, and at a collector in the scattering chamber itself. The James Frank Institute had a unique ability to safely liquefy and use liquid hydrogen, thanks to Professor Lothar Meyer, who had designed a liquefier for hydrogen, improving on the equipment he had used in Leiden, and Mr. Ray Szara, cryogenic engineer and a former president of the Cryogenic Society of America. We used the more dangerous but 20 times cheaper liquid hydrogen instead of liquid helium.

Because published work for designing skimmed supersonic beams was insufficient at the time, we studied the target so that we could optimize it for our purposes by varying, among other parameters, the nozzle-to-skimmer distance and the size and shape of the skimmer. As a part of his Ph.D. research work, Clifford Detz (University of Chicago Ph.D. thesis, December 1970) measured our jet's properties as the operational parameters were changed. He determined that the internal temperature of a pure argon beam in the jet could be as low as about 3 K and that in a pure nitrogen jet, 89–99% of its rotational energy was converted to translational energy. By mass spectrometric analysis he found that dimerization was less than about 0.1%.

We also had built a "universal" molecular beam vacuum chamber module. Our "universal" module was in the form of a cube, a stainless steel walled chamber about 18 inches on a side. This provided six faces for standard flanges that could support a beam source, a photocell, a vacuum pump, or a high-voltage or rf feed-through, which could be set up and taken down very quickly.

In 1972–1973 I took an unconventional sabbatical in industry as a consultant and then as vice president of engineering in the electric power equipment manufacturing business, at the ITE Imperial Corp. of Philadelphia. This sabbatical arose from my interest in high-voltage apparatus for electric deflection and acceleration of molecules. When I returned to the University of Chicago in 1973, my industrial experience had impressed me with the benefits of collaboration that I had not appreciated before. I purposely sought out my university colleagues to collaborate to an extent that I never had before.

Collaboration proved providential. Rick Smalley had recently come to the University of Chicago in 1973 as a postdoc. He and Donald Levy were interested in "solving" the $NO_2$ visible spectrum problem. They came to my office to discuss the feasibility of simplifying the $NO_2$ visible spectrum by selecting out the low rotational states in a molecular beam, using [our] electric quadrupole rotational state selectors, and measuring laser-induced fluorescence from it. I suggested that the rotational partition function would be too large to give a useful density of low rotational states from a state selector, and that a better idea would be to rotationally cool the $NO_2$ in a supersonic jet. Based on our measurements on Clifford Detz's jet source, one would not need state selectors at all. Best would be seeding the $NO_2$ in a supersonic argon monatomic gas jet. As I recall, Rick and Don were skeptical even though I "guaranteed" [based on Detz's measurements] that the rotational temperature would be below 15 K. After all, the argon's translational temperature was about 3 K and rotational relaxation was nearly complete for nitrogen, and there was little polymerization. Daniel Auerbach, also at Chicago with us as a postdoc, was an important contributor to the discussions we had. I personally felt so certain this approach to the $NO_2$ problem would work that I volunteered to supply the jet: I had just the apparatus to do it—Clifford Detz's target, which could be fitted into the "universal" module. Instead of scattering lithium atoms off the jet, we would scatter light. It would be just a matter of a few of weeks to modify the target to fit on to the "universal module" and to test out the idea.

It was a matter of weeks to set up a trial run with an argon ion laser. The experiment "worked the first time." The argon ion laser was tunable over a sufficient frequency range to observe many lines. Comparing the stimulated fluorescence spectra of a room temperature $NO_2$ sample, a neat $NO_2$ jet and an argon jet seeded with $NO_2$, individual rovibronic lines were only resolvable in the seeded supersonic jet. The $NO_2$ had reached an estimated rotational temperature of 3 K. Soon Rick Smalley and Don Levy set up a tunable dye laser to scan more broadly over the visible spectrum. The results were published by R. E. Smalley, B. L. Ramakrishna, D. H. Levy and myself in November 1974.

Enrico Fermi reportedly used to say, "Research is 85% procurement."

As one can imagine, it was fairly time-consuming, painstaking and expensive to use liquid hydrogen for the $NO_2$ work. In teaching myself about molecular beam sources, in the early 1960s I had attended the 4th Rarified Gas Dynamics Symposium in Toronto in 1964 where a young French scientist, Roger Campargue, had presented a paper that claimed that you did not need huge diffusion pumping capacity to create a very cold supersonic jet. Campargue's work at the time seemed too good to be true. About a decade later, after our work with the argon seeded jet, I went sailing with Peter Toennies at a Gordon Conference in New Hampshire. Toennies told me that he had tested a helium Campargue jet and it had worked beautifully. Shortly thereafter I happened to be in Hannover in Germany and had a free weekend. I had the idea to telephone Roger Campargue from Hannover to ask if I could visit with him on a Saturday in Paris. My rusty French enabled me to penetrate the Saclay switchboard in Paris and reach Roger. Cautiously he said that he would meet me at my hotel in Paris in the afternoon. That afternoon he was most gracious and generous, explaining exactly what principles to follow to make the Campargue source work. He invited me to dinner at his apartment that evening in Paris, and I remember a delicious meal and the great pleasure of meeting his charming wife. Incidentally Campargue explained that his Toronto presentation had been made deliberately vague because the technique at the time was deemed to be of strategic importance for uranium isotope separation.

Upon my returning from Paris we set up a Campargue 100 bar helium source in our "universal" module. The Low Temperature Laboratory of the James Franck Institute already had low-temperature physics utility stations that Lothar Meyer, Ray Szara and some of the low-temperature physicists at the James Franck Institute had set up. They provided high-capacity vacuum service that could operate in the micron pressure range for pumping liquid helium. Their vacuum stations used Stokes® ring jet booster pumps and large Kinney® mechanical pumps to lower the pressure and hence boiling point temperature of evaporating liquid helium in cryostats. One of these vacuum stations became the vacuum system for our Campargue source.

Incidentally the Stokes ring jet booster pumps used polychlorinated biphenyl PCB pump oil. Only PCBs and asbestos share "the notoriety of being specifically regulated under the US Toxic Substances Control Act, section 6(e) of TSCA, 15 U.S.C. §2605(e) (1.)."

It is interesting that our Campargue supersonic helium jets could reach a lower internal temperature than the 1 K of vacuum-evaporating liquid ⁴He in the James Frank Institute's low-temperature physics cryostats.

I did not have unrestricted research money to pay for the modifications and the operational liquid hydrogen needed for adapting and running the supersonic jets for the $NO_2$ experiments. For the first time in my research career, I spent money that I did not have—to make the modifications and pay for the cryogenic liquids. By the time we had set up the Campargue free jet work, I had spent about $30,000 that was not mine, an amount that fortunately was covered by James Frank Institute discretionary funds, after the fact, thanks to its director, Professor Ole Kleppa.

September 24, 2012

# A2

# Up the carbon path

## David E. H. Jones
*Department of Chemistry, University of Newcastle upon Tyne, NE1 7RU, UK
davidjones9@googlemail.com

My musings about carbon probably started with annoyance. Like many motorists, I found that the process of "decarbonizing" an engine meant taking it all apart and scraping the carbon off each part. Solid carbon does not have a solvent which you can just flush through an engine to dissolve the carbon away. Indeed, very few solid elements are soluble in anything. Their atoms combine into molecules too big to dissolve in any solvent. According to old textbooks, carbon atoms can form graphite or diamond, both big insoluble polymers. I have mused about the art of splitting diamonds, thus creating two new faces. Each face must rapidly saturate its newly exposed carbon valencies by combining somehow with the oxygen or the water vapour of the air.

---

*Former guest member of staff.

---

$C_{60}$ *Buckminsterfullerene: Some Inside Stories*
Edited by Harry Kroto
Copyright © 2015 Pan Stanford Publishing Pte. Ltd.
ISBN 978-981-4463-71-3 (Hardcover), 978-981-4463-72-0 (eBook)
www.panstanford.com

Meanwhile, in my role as "Daedalus" of the magazine *New Scientist*, I had to invent a column of scientific speculation every week. In the column of November 3, 1966, I began to muse on the strange difference of density between solids (typically a few grams per cubic centimeter) and gases (typically a few milligrams per cubic centimeter). A hollow molecule might be somehow intermediate. I began to muse about carbon, and was aided by D'Arcy Thompson's exposition in *On Growth and Form* of the deep-sea radiolarian *Aulonia hexagona*. Under the microscope, this is a tiny, closed, spherical surface essentially made of hexagons, each composed of six silica rods about 0.01 mm long. I imagined a hollow carbon molecule, essentially a closed-shell graphite sheet that imitated Aulonia. It would have about 1200 carbon atoms. I knew that a molecule which formed a spherical shell and enclosed space had to have exactly 12 pentagons somewhere among its hexagons. The mathematician Euler had demonstrated that inevitable 12.

When I included that column in my book *The Inventions of Daedalus* (W. H. Freeman, 1982), I carelessly attributed the "12" insight not to Euler but Thompson. I made many calculations about hollow carbon molecules, but failed to imagine the smallest one possible—which is of course the $C_{60}$ buckminsterfullerene. I implied that a material made of large hollow molecules would be supercritical in ambient conditions. To my great surprise, $C_{60}$ buckminsterfullerene was later made. It was a conventional solid; but at least it was soluble in numerous solvents.

# A3

# In my time: scenes of scientific life (extract)

## Sydney Leach

*Annu. Rev. Phys. Chem. 1997. 48:1–41*
*Copyright © 1997 by Annual Reviews Inc. All rights reserved*

### IN MY TIME: Scenes of Scientific Life

*Sydney Leach*
Département Atomes et Molécules en Astrophysique, CNRS-URA 812, Observatoire
de Paris-Meudon, 5 Place Jules-Janssen, 92195-Meudon, France;
e-mail: leach@obspm.fr

Another unusual spin-off resulted from the work on electron impact spectroscopy. At the beginning of the 1960s I was a once-a-fortnight consultant to the Chemical Physics section of the Commissariat à l'Energie Atomique (CEA) at Saclay. This section was headed by the French Alsatian Herman Hering, a man of great rigor who had survived a German concentration camp, and later by the excellent British chemist Jack Sutton. Since they had plenty of funding, I suggested that their new young scientist, Roger Campargue, should build a true molecular beam of the Kantrowitz & Grey type (93) and cross it with a controlled electron beam so as to do improved spectroscopic experiments, on cold supersonic molecular beams, with respect to my Orsay effusion devices. Campargue built the molecular beam but, to my increasing frustration, never did the crossed beam experiment. Instead he spent several years studying the thermodynamic properties of his molecular beam. He was quite right to do so, since it enabled him to build more compact and powerful supersonic molecular beams, of the free jet variety, requiring relatively small pumps (94a,b). Beams were in the French tradition, since the originator of atomic beams was Louis Dunoyer, who, in 1911, had made a beam of sodium atoms (95a,b). The Campargue technique was adopted by many of the early molecular beamists; in fact, the crossed electron-molecular beam experiment of my dream was first done by Don Levy in Chicago using a Campargue-type beam (96) and by Terry Miller at IBM (96). Many years later Roger Campargue kindly gave me his original apparatus, which was then used at the Laboratoire de Photophysique Moléculaire by Benoit Soep in his initial molecular beam experiments (97).

93. Kantrowitz A, Grey J. 1951. *Rev. Sci. In-strum.* 22:328
94a. Campargue R. 1964. *Rev. Sci. Instrum.* 35:111
94b. Campargue R. 1984. *J. Phys. Chem.* 88: 4466
95a. Dunoyer L. 1911. *C. R. Acad. Sci.* 152: 594
95b. Dunoyer L. 1911. *Le Radium* 8:142
96. DeKoven BM, Levy DH, Harris HH, Zagarski BR, Miller TA, 1981. *J. Chem. Phys.* 74:5659
97. Campargue R, Soep B. 1979. *Chem. Phys. Lett.* 64:469

Dear Harry,

My only further remarks on the Buckyball story are perhaps only of indirect interest. They concern an early use of the laser vaporisation technique. Following a 1963 paper by Boris Stoicheff [1] in which he attacked solid targets with a ruby laser, I started to work in this area, calling my technique "Laser Pyrolysis." After building a ruby laser, my initial idea was to have the plasma formed by laser pyrolysis (we studied 22 different elements as targets) react with different ambient gases so as to form new emitting molecules and ions [2, 3], expecting to observe new spectra. This was extended to the production of inverted populations of electronic levels in Al IV (which eventually led later to X-ray laser emission) [4], and to the use of the laser pyrolysis technique to measure radiative lifetimes of the C2 molecule [5, 6]. For this latter work we developed a device so that the cylindrical graphite target could be helicoidally rotated so that a fresh target surface was presented periodically to the laser beam. Someone mentioned later, in a publication I can't recall, that this was the first time a device of this kind was used in laser vaporisation studies.

At this point in time in the early part of the 1970s, we at Orsay were able to persuade the high-energy physicists to drill holes in their storage ring bending magnets and so my interests switched to using synchrotron radiation and I dropped the laser pyrolysis work. Pierre Jaegle continued the laser pyrolysis work in developing VUV and X-ray emission sources in which he contributed significantly to the development of EUV lasers [7].

1. Proc. Xth Colloquium Spectroscopium Internat., ed. E. R. Lipincott, M. Margoshes, Spartan Books, Washington D.C., 1963, p. 399.

2. B. Ban, S. Leach, G. Taieb, M. Velghe, *J. Chimie Physique* **64** (1963) 397–400.

3. A. Frad, S. Leach, *Chem. Phys. Lett.* **12** (1972) 599–601.

4. P. Dhez, P. Jaegle, S. Leach, M. Velghe, *J. Appl. Phys.* **40** (1969) 2545–2548

5. M. Velghe, S. Leach, *J. de Physique* **34** (1973) C2-111–116.

6. S. Leach, M. Velghe, *J. Quant. Spectrosc. Radiat. Transfer* **16** (1976) 861–871.

7. G. Jamelot, in *Physics with Multiply Charged Ions*, ed. D Liesen, Plenum, New York, 1995, pp. 291–317.

I don't know if any of this is useful for your historical efforts.

Kindest regards,
Sydney

# Early days in the Rick Smalley lab

## Michael A. Duncan

*Department of Chemistry, University of Georgia, Athens, GA 30602, USA*

maduncan@uga.edu

I joined the Smalley lab at Rice University in the fall of 1977, as a first-year graduate student in Physical Chemistry. Smalley had only arrived there himself the year before, so the group was young and in the process of building up new equipment. Two graduate students were in the group who had started the year before, Greg Liverman and Steve Beck. About this time, Dr. David Monts joined the group as a postdoc. Tom Dietz, another first-year graduate student, joined the group at the same time that I did, and we became partners in the lab.

At that time, experiments were in progress using a small vacuum chamber known as apparatus #1, or "App 1" in the group lingo. This chamber had a continuous free-jet expansion and laser-induced fluorescence detection. The

$C_{60}$ *Buckminsterfullerene: Some Inside Stories*
Edited by Harry Kroto
Copyright © 2015 Pan Stanford Publishing Pte. Ltd.
ISBN 978-981-4463-71-3 (Hardcover), 978-981-4463-72-0 (eBook)
www.panstanford.com

laser used was a nitrogen laser-pumped homemade dye laser.

The main chamber for App 2 was already constructed, but no experiments were yet in progress there. Dietz and I were given the job of completing App 2 according to designs developed by Smalley. The plan was for the large main chamber (1.0 meter diameter cylinder, 1.0 meter high) with a skimmer connecting into a series of three smaller chambers, each about one foot cubed. The beam was skimmed again in chamber #2, chamber #3 would contain the mass spectrometer, and chamber #4 would catch the gas at the end of the line to pump it away. An early figure of App 2 from my Ph.D. thesis is shown in **Fig. 1**, in which the last chamber in the line is not shown, and the third chamber has the flight tube for the mass spectrometer mounted in the vertical direction.[1]

Dietz and I were given our own construction projects. Each of us would take the lead on a topic, but we continued to work together on most everything. I was given the job of designing and building a time-of-flight mass spectrometer for the downstream detection of the molecular beam to be installed in chamber #3. I had very little training in electronics, and none in the construction of mass spectrometers, so I had to start from scratch. I got lots of help from the departmental electronics technician, who had worked previously for Bendix Corporation making mass spectrometers. He directed me to the classic 1955 paper by Wiley and McLaren,[2] which I referred to in beginning the job. The instrument shop at Rice was also extremely helpful in the design stage of the work.

Smalley's original motivation for having a mass spectrometer was the study of photodissociation dynamics, particularly for the system of formaldehyde. At that time, formaldehyde photodissociation in the $n \rightarrow \pi^*$ system was understood to involve either a radical (1) or molecular

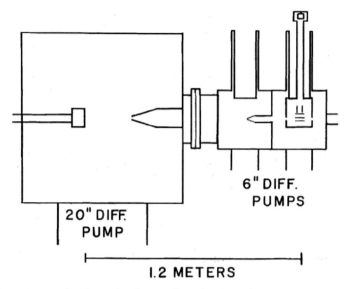

**Figure 1** An early schematic of App 2 from the Ph.D. thesis of M. A. Duncan.

process (2). Key experiments had been done by Paul Houston and Brad Moore at UC-Berkeley.[3] The branching ratio between these was hotly debated because of the difficulty in detecting the photofragments.

$$CH_2O + h\nu \rightarrow CHO\cdot + H\cdot \tag{1}$$

$$CH_2O + h\nu \rightarrow CO + H_2 \tag{2}$$

Smalley had the idea that we could detect these fragments with mass spectrometry. To do this, we designed a time-of-flight instrument with a dual ionization source **(Fig. 2)**. The normal configuration would detect molecules in the molecular beam, while the photofragment configuration was just offset from the molecular beam axis and would detect molecules ejected out of the beam by photodissociation. In both regions, ionization would be accomplished with electron impact using a heated

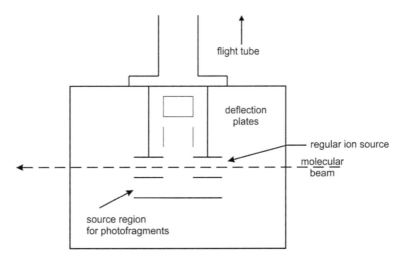

**Figure 2** Schematic of the double ion source time-of-flight mass spectrometer built for photofragment detection.

wire filament to generate the electron beam. The mass spectrometer was designed and built by early spring 1978, and initial experiments with it started. In the beginning, the electron gun and the voltages for the ion acceleration plates had to be pulsed. I designed and built transistor-based circuits to be triggered with a commercial pulse generator and then stepped up the voltages to the required levels. Because I knew virtually no electronics, the first circuit attempts went up in flames (literally). Likewise, we had trouble running the filaments for the electron beam, as these had the tendency to move upon heating and then to short to nearby grounded plates. After several iterations, both problems were solved and we obtained our first mass spectrum on background gas on April 1, 1978. A sketch of this from my lab notebook is shown in **Fig. 3**. At this time we had no way to digitize a mass spectrum; we sketched the early ones, then used an oscilloscope camera to photograph these, before eventually getting a transient digitizer set up. Digital oscilloscopes came much later.

Dietz was given the job of putting together a homemade dye laser using our new Nd:YAG laser which arrived at about this time. It was a Quanta Ray laser (serial number 3), the company which was later bought by Spectra Physics. Gene Watson, the company owner, came to install this for us. I remember us putting together the first version of the dye laser using the grazing incidence design published about that time by Littman.[4] We did it on Super Bowl Sunday in February 1979, with a TV set up on the laser table so we could also watch the game.

The nozzle for App 2 was to be a pulsed valve, designed after the concepts developed by Ron Gentry at Minnesota.[5] Gentry was already using a spring-steel current loop designed nozzle, optimized for a short pulse duration (10–20 μs). This valve was sold commercially for many years.

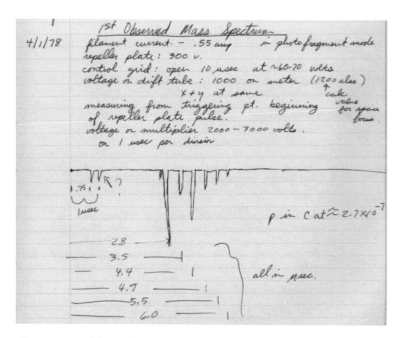

**Figure 3**  MAD lab notebook sketch of the first mass spectrum measured on App 2.

Smalley had the idea that the same design could be used with a longer spring to make a nozzle with a wider pulse in time (200–300 µs) to achieve better supersonic cooling and complex formation. Greg Liverman had already made such a nozzle, and it was already working when Dietz and I arrived. We joined in with its testing, and eventually took it over as part of App 2. Tests of its performance were done in summer 1979 with a nude ionization gauge mounted in chamber 2, measuring beam intensity as the nozzle moved in and out with respect to the skimmer, with more or less backing pressure. We learned that nozzle–skimmer interactions are important, that an optimized system had to have a sharp skimmer spaced away from the apparatus wall to avoid reflected shocks, and that the nozzle must also stay away from the skimmer enough to avoid shock waves from scattering off the skimmer. We also learned that the long spring did give a longer pulse of gas, but that it bounced and that many beam pulses were doubled or tripled from this, as shown in the oscilloscope trace of the beam profile in **Fig. 4**. The operation of this valve was described in an early publication from the group.[6] Unfortunately, the spring-steel design for this nozzle was highly problematic. It involved deposition of an insulating film of Torr-Seal epoxy on the nozzle flange, and these films cracked over time, shorting the spring to ground. The lifetimes of these valves could vary between a few weeks to a few months before they self-destructed. Much time was lost rebuilding pulsed valves. About 1981 (after I and Dietz had graduated and left the group) Ed Grant at Cornell published a design for a double solenoid valve,[7] which later was sold commercially by the Newport Corporation. Smalley liked this design and switched over to it, modifying it to his own tastes. This double solenoid was used from that time on for all experiments and was found to cool molecules as well as the earlier design, but to be much more reliable. The double solenoid valve was used for later fullerene experiments.

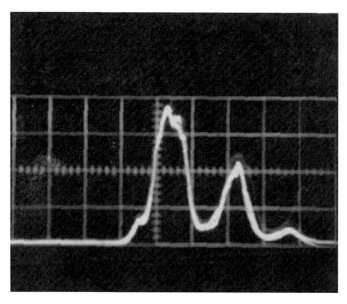

**Figure 4** Oscilloscope photo of the profile of the molecular beam pulse measured with a nude ionization gauge, with two "bounce" pulses. The horizontal axis was 200 microsecond/division.

We eventually tried an experiment with formaldehyde, and were able to detect its mass spectrum in the molecular beam. However, when we tried to activate the fragmentation detection, using a dye laser tuned near 280 nm, we were unable to detect any photofragments. We decided that the problem was the weak $n \rightarrow \pi^*$ transition for formaldehyde (it is a forbidden electronic transition, becoming weakly allowed by vibronic coupling), the low vapor pressure that we were able to achieve by heating the solid paraformaldehyde, and its low quantum yield for photodissociation. To improve matters, I decided to try a molecule with a much stronger transition, much greater vapor pressure, and unit quantum yield. I settled on iron pentacarbonyl, $Fe(CO)_5$. This decision turned out to be our first lucky accident.

When we used iron carbonyl in the molecular beam (June 25, 1979), we immediately found its mass spectrum with electron impact ionization, and the signals were strong. We then changed over to the fragment detection configuration of the machine, and turned on the dye laser, again near 280 nm. We detected photofragments, and were initially pleased that this part of the experiment was now working. But we had difficulty with an enormous spike that kept moving across our mass spectrum. We first thought that this was more trouble with my bad electronics, but then we noticed that it moved in time with the laser firing. Finally, we noticed that it went away when the laser was blocked or when the ion detector was turned off. By these steps, we realized that we were seeing ions produced by the laser! We changed the oscilloscope triggering to start on the laser instead of on the acceleration plate switching, and were then able to make sense of the spectrum. The gigantic spike was $Fe^+$ produced from $Fe(CO)_5$ by a multiphoton absorption and fragmentation process. This led to our first paper on photoionization, as we compared this behavior for several transition metal carbonyls.[8] We compared the signals generated from electron impact and photoionization, and realized that the photoionization signal was about 1000 times larger. At this point Smalley remembered seeing research articles from the Schumacher group in Switzerland, who used resonance-enhanced photoionization to detect alkali metal dimers.[9] We then made the decision to abandon formaldehyde and also my electron impact ionization rig and pulsing circuits, and to move forward in the completely new direction of photoionization detection. This made the mass spectrometer much simpler to use, since no filaments were required and no plates needed to pulse. We applied DC acceleration voltages directly from power supplies, and the dye laser pulsing provided the starting time for the time-of-flight measurement.

The next year or so was spent investigating the spectroscopy of various small molecules using resonant two-photon photoionization (R2PI). In this process, dye laser excitation produces excited state molecules with the first photon absorbed, and then absorption of a second photon (from the same laser at the same wavelength) produces an ion. This works only if the excited state investigated lies more than halfway to the ionization potential, but this turns out to be true for many medium-sized organic molecules. We studied bromobenzene, benzene, toluene, aniline, biacetyl, naphthalene, and other related species. In addition to single-color, two-photon experiments, we also used two dye lasers for two-color experiments. Later on, the group used this method for studies of many metal dimers like Schumacher had done earlier, but the Smalley group was able to do transition metals using the laser vaporization source (see below). In addition to the ionization experiments, other experiments in the group continued to use laser-induced fluorescence (LIF) detection. David Powers

**Figure 5** The Smalley group in 1979, with App 2 in the background. L–R: Greg Liverman, Rick Smalley, Steve Beck, Mike Duncan, and Tom Dietz.

and John Hopkins had joined the group in the fall of 1979, and they worked on App 1 using fluorescence detection to probe a number of substituted benzenes, investigating intramolecular vibrational relaxation (IVR) using dispersed fluorescence measurements. A photo of the group in the lab about that time is shown in **Fig. 5**. **Figure 6** shows another photo of group members just about to leave on a trip to the Ohio State Spectroscopy Symposium in 1980, which was our first conference.

In about 1979, we developed a collaboration with the research group of Andy Kaldor and Don Cox at Exxon Research Labs in New Jersey. They were interested in infrared multiphoton photodissociation of uranium complexes in the gas phase as a method of isotope separation. The uranium complexes they were using had to be heated to get enough vapor pressure, and then their spectra were broad, with overlapping lines for the different isotopes. They wanted us to use the cooling of our pulsed nozzle

**Figure 6** The group members just about to depart for the June 1980 Ohio State Spectroscopy Symposium. L–R: Mike Duncan, Tom Dietz, David Powers, and Steve Beck.

expansions to improve this. We did the work, and it was successful. As part of our pay-off for beam time used, Exxon bought us one of the new UV rare gas-halide excimer lasers that were just coming onto the market. It was a Lumonics model, which could produce either 193 nm (ArF) or 157 nm ($F_2$). We soon found that this laser could produce ionization for many of our organic molecules and their clusters in the beam via resonant two-photon ionization through highly excited states. This excimer laser turned out to be another critical part of the puzzle for later work. Another part of our collaboration with Exxon was the construction of App 3, which was smaller than App 2 but with the same capability, and its copy which was sent to Exxon in New Jersey. The App 3 copy was used later by Exxon in their studies of carbon clusters.

One particular experiment using LIF detection was designed to obtain the spectrum of the phenyl radical. We used iodobenzene, and photolyzed it with a UV laser crossing the expansion, hoping to produce the desired radicals and to cool them. When we scanned the electronic spectrum, we did not find the phenyl radical. Instead, we found by accident a beautiful LIF spectrum of the $I_2$ molecule. This observation showed us that the collisions in the early phases of the supersonic expansion could be effective in promoting *recombination reactions*. This was of course already known in the molecular beam community, as three-body recombination was recognized to be responsible for growing clusters. But the production of such an intense signal of $I_2$ from the photolysis of iodobenzene made a strong impression on us.

At this time there was significant interest in the Physical Chemistry community in small metal clusters. Metal dimers and trimers were being produced in rare gas matrices by Bill Weltner at Florida, Martin Moscovits at Toronto, and several others. Jim Gole at Georgia Tech had made a heated oven source with temperatures high enough to produce copper dimer and trimer in molecular beams.[10] In late 1980, Smalley became interested in this also. He noted our result on iron pentacarbonyl, in which photolysis produced neutral iron atoms, and then ionized them. He saw how efficient the recombination of iodine atoms was in the iodobenzene experiment, and he got the idea that we could photolyze $Fe(CO)_5$ early in the expansion and then the iron atoms formed could recombine to grow iron atom dimers, trimers, etc.

We therefore set up an experiment to photolyze $Fe(CO)_5$ with our Nd:YAG laser at 532 nm crossing the molecular beam expansion just a few millimeters downstream from the nozzle opening. We set up the ArF excimer laser crossing the molecular beam in the mass spectrometer to ionize the iron atom clusters that might be produced. Initially, we saw no signal at all for iron atoms. After discussing why this might be so, we decided that the photolysis laser overlap with the molecular beam might not be optimum. We decided to bring this laser down the molecular beam axis, counter-propagating to the beam direction, but collinear with it. This was quite difficult to do because of the long distance between the last window in the third small chamber and the nozzle (nearly 2.0 meters), but after some effort we had the beam aligned and started the experiment again. After some scanning of the YAG and excimer laser timings relative to the nozzle pulse, we eventually found a signal consisting of several mass peaks. However, when we tried to assign the masses, we found that there was no iron. Instead, there were peaks assigned to multiples of mass 27 and adducts of these with acetone. The acetone made sense because we had recently rinsed the gas lines with it. But mass 27 was *aluminum* and not iron! We realized in a few minutes that the photolysis laser had been focused to excite the gas coming out of the nozzle, and that it had hit the front plate of the nozzle just beside the gas outlet.

We had vaporized the nozzle plate, made of aluminum, and the metal vapor had recombined in the expansion (as planned) producing aluminum clusters! This completely accidental discovery was how the laser vaporization source was "invented."

When we realized what had happened, we decided to test the idea by mounting pieces of other metals beside the nozzle, in the attempt to form clusters of other metals. Unfortunately, this required more precise alignment of the vaporization laser, and we could not see into the main chamber easily because there was no window along the molecular beam direction other than the one that the laser was going through. After discussion, we decided that we needed a window a few degrees off the main molecular beam direction to look through to see the laser hitting the target. This discussion took place in the middle of an experiment, with all the molecular beam chambers under vacuum. After an animated discussion, including Smalley as a participant, he was inspired by the idea, and went immediately to the tool chest and grabbed a power drill and bits. As we all watched in horror, he drilled right into the wall of the App 2 main chamber, making a hole for the desired window, while the experiment was still under vacuum! He immediately sanded down the surface around this, got a thick O-ring and window, and pressed the window over the O-ring around the hole. The vacuum (and some tape) held the window in place, and so we could continue the experiment. The window worked as desired and we were able to align the vaporization laser to hit the target. We rapidly submitted a communication on our aluminum cluster discovery.[11] **Figure 7** shows a photo of Smalley aligning the excimer laser used for photoionization of the metal clusters made about this time. In the lower right-hand corner of the photo, the laser alignment window can be seen, still held in place with tape!

**Figure 7** Smalley lining up the App 2 excimer laser. Note laser alignment viewport hole with tape at lower right.

We soon found that the laser vaporization method was completely general, and could be used for many metals. In initial experiments, we realized that the laser would drill holes in metal plugs after only a few minutes, and thus we discussed the need for a continuously refreshed metal surface. Smalley had the idea of the motor-driven rotating rod source, with the laser coming in from the side, but I remember thinking that this would be way too complicated to work in a real experiment (I was of course wrong). A design was placed in the shop for this, but this all happened at about the time that Tom Dietz and I were scheduled to have our thesis defenses and to graduate (in August 1981). I had a postdoc lined up with Steve Leone in Boulder, CO, and Tom had a job lined up with Exxon in New Jersey, and so we both had to move on to other things. Mike Geusic had arrived in the meantime as a new grad student, and both Christina Puiu and Pat Langridge-Smith had arrived as new postdocs, and so we left the continued development of the source and its new experiments with Powers, Hopkins, and these new people. The first new paper using the source was an electronic spectrum of copper dimer,[12] and this paper provided more information about the rotating rod source design. Other experiments soon after that included studies of metal cluster reactions and the production of semiconductor clusters.

Only later after our communication on the source appeared in the literature did we realize that Vladimir Bondybey at AT&T Bell Labs had been working along similar lines completely independently. He had apparently set up laser vaporization "on purpose" to vaporize lead dimers, which he detected with laser-induced fluorescence.[13] Bondybey's work happened about the same time as ours, but we could see more of what was going on with cluster formation because of the laser photoionization and mass spectrometry, which he did not have. The source thus became known as the Smalley source rather than the Bondybey source. More details about the laser vaporization method and how it has been used since this time are contained in a recent review on the subject.[14]

It is now clear that the essential ingredients for the cluster experiments were the supersonic beam, the pulsed nozzle and its incorporation into App 2, the photoionization detection (particularly with the excimer laser), the time-of-flight mass spectrometry, and then the laser vaporization method. Also critical were the arrival of the first new YAG and excimer lasers, both of which only became available commercially at this time. These instrumental developments were all used later in the discovery of $C_{60}$.

### References

1. M A Duncan, "Resonant laser ionization in the study of molecular excited states," Ph.D. thesis, University Microfilms International, #8216315, Rice University, 1982.

2. W C Wiley and I H McLaren, "Time-of-flight mass spectrometer with improved resolution," *Rev Sci Instrum*, **26**, 1150 (1955).

3. P L Houston and C B Moore, "Formaldehyde photochemistry: appearance rate, vibrational relaxation, and energy redistribution of the CO product," *J Chem Phys*, **65**, 757 (1976).

4. M G Littman and H J Metcalf, "Spectrally narrow pulsed dye laser without beam expander," *Appl Optics,* **14**, 2224 (1978).

5. W R Gentry and C F Giese, "Ten microsecond pulsed molecular beam source and fast ionization detector," *Rev Sci Instrum*, **49**, 595 (1978).

6. M G Liverman, S M Beck, D L Monts, and R E Smalley, "Laser characterization of pulsed supersonic molecular jets and beams," *11th International Symposium on Rarified Gas Dynamics*, **2**, 1037 (1979).

7. T E Adams, B H Rockney, R J S Morrison, and E R Grant, "Convenient fast pulsed molecular beam valve," *Rev Sci Instrum*, **52**, 1469 (1981).

8. M A Duncan, T G Dietz, and R E Smalley, "Efficient multiphoton ionization of metal carbonyls cooled in a pulsed molecular beam," *Chem Phys*, **44**, 415 (1979).

9. A Hermann, S Leutwyler, E Schumacher, and L Wöste, "Multiphoton ionization: mass selective laser spectroscopy of $Na_2$ and $K_2$ in molecular beams," *Chem Phys Lett*, **52**, 418 (1977).

10. D R Preuss, S A Pace, and J L Gole, "The supersonic expansion of pure copper vapor," *J Chem Phys*, **71**, 3553 (1979).

11. T G Dietz, M A Duncan, D E Powers, and R E Smalley, "Laser production of supersonic metal cluster beams," *J Chem Phys*, **74**, 6511 (1981).

12. D E Powers, S G Hansen, M E Geusic, A C Puiu, J B Hopkins, T G Dietz, M A Duncan, P R R Langridge-Smith, and R E Smalley, "Supersonic metal cluster beams: laser photoionization studies of $Cu_2$," *J Phys Chem*, **86**, 2556 (1982).

13. V E Bondybey and J H English, "Laser induced fluorescence of metal clusters produced by laser vaporization: gas phase spectrum of lead ($Pb_2$)," *J Chem Phys*, **74**, 6978 (1981).

14. M A Duncan, "Laser vaporization cluster sources," *Rev Sci Instrum*, **83**, 041101 (2012).

# A5

# Discovery of IRC+10216[*]

## Eric Becklin

UCLA Physics & Astronomy, 475 Portola Plaza, Los Angeles,
CA 90095-1547, USA
becklin@astro.ucla.edu

IRC+10216 was first picked up on the Two Micron Sky Survey of Gerry Neugebauer and Bob Leighton in late 1964. The survey was carried out at Mt. Wilson using a 62-inch epoxy telescope built by Leighton and Neugebauer in the Physics labs at Caltech. The source was not noticed as exceptionally red in the 0.8 to 2.2 micron color of the survey because of source confusion at 0.8 micron of a nearby blue star, but no one knew of or expected this confusion.

I had worked on the sky survey as Neugebauer's graduate student, so in 1968 and early 1969 decided that it was probably worth looking at the reddest sources in the two micron catalog at other wavelengths, particularly at longer wavelengths of 3.5 and 10 microns. With support from Neugebauer, I put together a list of about 50 of the reddest sources in the northern sky; the list included IRC+10216.

[*]Dedicated in memory of Gerry Neugebauer (1932–2014).

$C_{60}$ Buckminsterfullerene: Some Inside Stories
Edited by Harry Kroto
Copyright © 2015 Pan Stanford Publishing Pte. Ltd.
ISBN 978-981-4463-71-3 (Hardcover), 978-981-4463-72-0 (eBook)
www.panstanford.com

With a Caltech graduate student, I went to the old Mt. Wilson 60-inch telescope to make observations with a photometer that had been developed with Neugebauer for my thesis work. The 10 micron detector system was on loan from Jim Westphal in the Geology Department at Caltech. Most of the sources had 3.5 and 10 micron fluxes predicted from the survey fluxes. I was in fact considering stopping the follow-up because there was so little new coming out of the work. Then, in the middle of the night on 23 March 1969, all that changed. We came to a source that was about 10 times brighter at 10 microns than any other non-solar system source known in the Northern Hemisphere!! **Figure 1** shows a strip chart of the very first measurement at 10 microns. It was so exciting that, as I went to call Neugebauer on the phone, I started to hyperventilate. I knew that we had just discovered the most important source from the Two Micron Sky Survey.

Over the next several months, a number of us in the IR lab at Caltech started a major campaign to determine the nature of IRC+10216. Of particular importance in this work was Harry Hyland, a postdoc of Jesse Greenstein's who was working with us in the IR lab. From the photometry we determined that it was radiating with a temperature of about 600 K. The survey history showed a variable source at 2.2 microns with an amplitude change of a factor of over 5 and with a period of about 600 days. Hyland and Jay Frogel took low-resolution spectra in the 2 micron region and determined that there were, at most, very weak absorption or emission features, much less than seen in the other red sources found in the survey or known from long period variable stars such as Mira. Equally interesting, H. C. Arp of the Carnegie Institution of Washington, obtained a 200-inch photographic image that showed an elliptical source of about 4 × 6 arcsec. Although the source was out of the Galactic plane, we rejected an extragalactic source. From all of this, we concluded that the object was a star losing so much mass that it was optically thick at 2.2 and

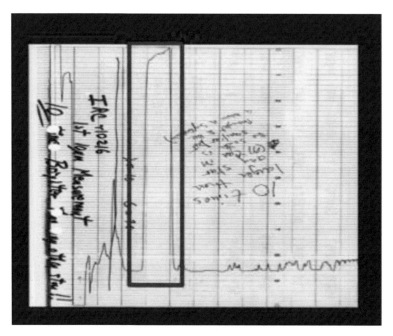

**Figure 1** The first measurement of the 10 micron flux from IRC+10216. Note that the source is almost off scale. This was almost 10 times larger than any previously measured star in the Northern Hemisphere!

maybe also at 10 microns. It was also relatively nearby with a galactic distance of about 200 pc.

I had been invited to give a talk on new infrared sources at the Gordon conference in June 1969. Of course, I talked mainly about IRC+10216. Because it was a chemistry conference, I speculated that IRC+10216 might become important in our understanding of mass loss and chemistry from stars. Fred Whipple, who was in the audience and had just recently hired me at the Smithsonian, noted that this was a major discovery in astrophysics. A recent ADS survey shows about 1000 papers on IRC+10216, mostly in the area of astro-chemistry. The connection of IRC+10216 with the discovery of $C_{60}$ through Harry Kroto's work is especially exciting and an honor for me.

January 2013

# A6

# The discovery of the fullerenes

**Sean C. O'Brien**
*Rice Quantum Institute and Department of Chemistry, Rice University,*
*Houston, Texas 77251, USA*
c60chemist@yahoo.com

It is a fact of this universe that minor decisions and trivial details have vast repercussions. Seemingly meaningless details can years later drive apparently unrelated areas towards dramatically different conclusions. The Butterfly Effect applies to many surprising areas. Consider the following sequence of events:

In early 1983, my junior year at the University of Illinois, I needed to choose a senior thesis research project. I asked my advisor Dana Dlott about working with him and he recommended that I talk with Jim Lisy. Jim had recently begun planning some work on one of the most impressive tools in all of physical chemistry. Willis Flygare had built the magnificent Fourier transform microwave spectrometer. Lisy asked me to learn how to run the tool and then look for rotational transitions in van der Waals complexes of Ar and HF.[1] This was my introduction to supersonic

*$C_{60}$ Buckminsterfullerene: Some Inside Stories*
Edited by Harry Kroto
Copyright © 2015 Pan Stanford Publishing Pte. Ltd.
ISBN 978-981-4463-71-3 (Hardcover), 978-981-4463-72-0 (eBook)
www.panstanford.com

nozzles, isentropic cooling, cluster formation, and gigantic diffusion pumps.

That fall I was applying to graduate school. I wanted to go to either Berkeley or MIT. Prof. Lisy recommended that I apply to a third school and I took his advice. Thinking about wind chill factors I asked him if there were any good schools in Texas and he replied, "Rice is pretty good." So I applied to Rice, was accepted, and started graduate school in 1984. I had barely heard of Smalley before then, and for weeks after my decision, all the grad students at Illinois kept asking me, "Are you going to work for Smalley?"

Rice frequently accepted new graduate students for summer research preceding their first semester. Rick Smalley called me and invited me to work in his lab in the summer of 1984. That summer I worked on the downstream reactor nozzle, attaching reactant species to metal clusters. I earned a spot as author on two papers that summer.[2–3]

After careful study of a few different Rice professors I decided that I wanted to work for Bob Curl for my thesis research. When I asked him to join his group he said the only project he had available was on a semiconductor cluster grant with Smalley. Realizing that a Curl student would be a second-class citizen working in the Smalley lab I asked Rick if I could join his group. Had Bob accepted me into his basement laboratory I would never have worked in the Smalley lab on the third floor, never worked on clusters, and never designed the rotating disc source.

That year Rick Smalley, Bob Curl, and Frank Tittel were awarded a research grant from the US Air Force for the study of semiconductor clusters. Buried within this proposal was a commitment to study GaAs clusters. GaAs was an expensive material with only one commonly available form: a thin wafer. The Smalley cluster source required pencil-shaped rods of material. Rods of GaAs would have cost thousands of dollars even if they had been available, and they would have been extremely brittle. In order to meet the requirements of the research grant we needed to generate clusters from a disc of GaAs.

So we had to build a new cluster source. Andy Kaldor at Exxon had built a duplicate of the Smalley rod source and studied carbon clusters in 1984.[4] Their rod source was not capable of restricting the carbon cluster plasma long enough to facilitate extreme clustering conditions. This is probably the key reason they missed the discovery of fullerene chemistry; their source could not sustain the reactions necessary to allow the unreactive fullerenes to stand out, specifically buckyballs. Certainly their source made some buckyballs. But the ratios were not significant enough to highlight the special nature of $C_{60}$.

Harry Kroto had seen Rick Smalley's giant cluster generating chamber and decided that was where he could demonstrate reactions on small carbon chains. Harry initially failed to convince Rick to study carbon. We think one main reason is that Kaldor and Exxon had staked out their claim to carbon and Rick did not want to move into that territory. Had Harry been more persuasive he would have arrived months before the disc source was built, studied small carbon clusters made with the rod source (identical with Exxon's), and quite probably we never would have discovered the fullerenes. Fortunately for all of us, when Harry's irresistible persuasive powers collided head on with Rick's immovable wall, Rick won the first round delaying Harry's visit long enough for me to build the disc source.

## April/May/June 1985: The Rotating Disc Source

I asked to be assigned to design the new disc source and Rick approved the idea of my full-time effort on the source. During April and May I considered several different designs.

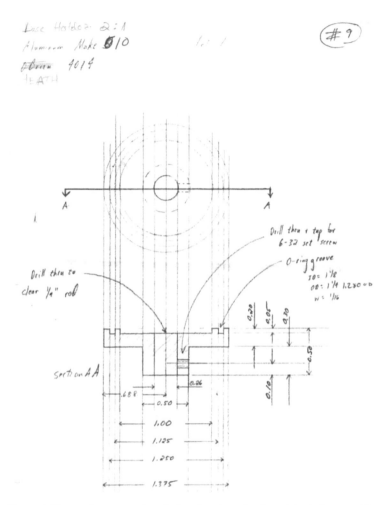

**Figure 1** Top and cross-section design drawings for the original disc cup. The quarter-shaped disc would be glued into the holder using an epoxy called Torr-Seal.

**Figures 1–3** show original design drawings for the disc holder, nozzle, and pivot arm. The disc was about the shape of a US 25 cent quarter.

Smalley drove the design of a pivot assembly. Phil Brucat, a postdoctoral research associate, added a key point

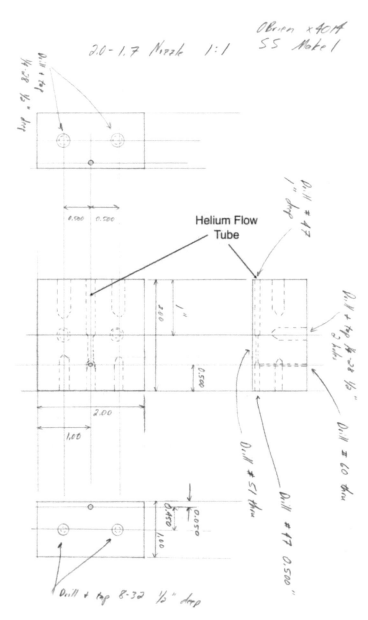

**Figure 2** Images of the nozzle block showing various orientations. The wall of the helium flow tube was extremely thin, but the Rice machine shop did a very good job of producing several of these.

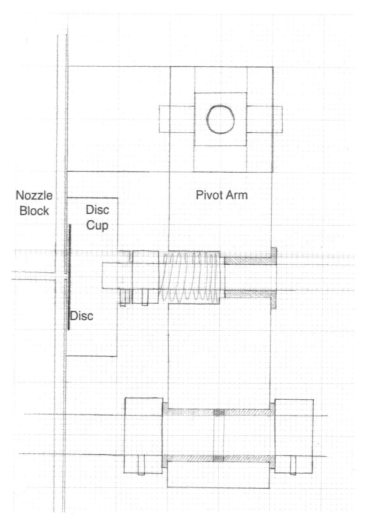

**Figure 3** Original design drawing cross-sectional view of the pivot arm, showing the target disc pressed up against the face of the nozzle block.

when he suggested a cam design for the motion of the pivot arm. But I believe the fundamental advantage of the disc source was my own design. I had designed a long cluster formation tube after the laser vaporization hole. These few millimeters of constrained gas flow led to dramatically higher cluster formation collisions. The rod source had a shorter confinement region, leading to many fewer clustering collisions prior to the supersonic expansion. The larger number of collisions allowed the fullerenes to grow larger and leave behind buckyballs.

The dimensions of the GaAs disc set a minimum path length for this constrained plasma. This path length is probably the fundamental reason this cluster source could produce the famous "flagpole" spectrum for $C_{60}$. Constraints such as this could have been added to the rod source had anyone wanted to do such a thing, but there was no reason to consider it until the results of September 1985.

By 1990 my disc source had been cloned in at least two labs: in Pat Langridge-Smith's lab in Edinburgh, Scotland, and a few miles south Harry Kroto had built another identical source for his cluster machine.

## August/September 1985

In 1985 there were no laser printers available. All our graphs for publications were hand drawn, traced by hand from the dot-matrix printouts. We would sit in Rick's office for hours at his drafting table, carefully drawing out the figures for our publications. One mistake, and we had to start over!

In early August I was sitting at the table minding my own business, carefully preparing the images for our GaAs paper,[5] when Rick came in and said, "Sean, guess who's coming? Harry!" I had no recollection of the brief conversation with Bob and Rick, so I asked, "Harry who?" He reminded me of Bob's friend who wanted to study carbon clusters. Little did any of us realize that Sean and Jim's days of studying semiconductor clusters were a nearly over. Our thesis research project was soon terminated and we would be carbon cluster researchers for the duration of our graduate school careers. Not one time was I asked about my preferred project. I was simply told to stop working on semiconductors.

A few days before Harry arrived we wanted to be sure the machine would work with carbon, so we inserted a graphite disc and took a mass spectrum with a "flagpole" $C_{60}/C_{62}$ ratio of about 25. Contrary to Mr. Baggott's version,[6] the reason we didn't notice the ratio was because the y-axis was set for the other fullerenes to be full scale. $C_{60}$ was well off scale but we had no idea how far. This is an important lesson for young scientists: make sure you look at all the data. Suppose the computer program had been set up for autoscaling? Maybe things would have gone a little differently. We might have started asking questions about $C_{60}$ two weeks before Harry arrived.

The events of the next two weeks have been well documented over the years and I have nothing significant to add to those accounts. I wish my memory were better and I could remember who first proposed that $C_{60}$ must be special or created the name buckminsterfullerene. But in the end those are just trivial details which undermine the unity of effort of this group.

My foray into the lab on Friday night, September 6, 1985, was a rare event; I truly wanted to contribute. When I shut off the machines that night I had the satisfaction that I had finally measured something new and that I was able to contribute real data to the group meetings without having to feel like a fifth wheel.

I now consider that a very important night. For the first time the ratio $C_{60}/C_{62}$ had been controlled and adjusted. The next morning Jim brushed aside my work as if it were trivial. His work that weekend with the integrating cup and backing pressure was very similar to adjusting laser power. They both control collisions and cooling and flight times down the molecular beam.

After the soccerball structure was discovered I was stunned at the speed with which the *Nature* paper was written.[7] To this day I consider writing a technical paper in two days extremely unprofessional. Science should be done carefully and thoughtfully. Writing a paper in 48 hours is simply absurd. And sure enough, we missed the credit owed to Harry Schultz, the first person to ever publish a truncated icosahedral $C_{60}H_{60}$ structure.[8] Several others, including Yoshida and Osawa,[9] Bochvar and Galpern,[10] Castells and Serratosa,[11] and Davidson,[12] had clearly discussed either $C_{60}$ or $C_{60}H_{60}$. The controversy surrounding the *Nature* paper would have been somewhat reduced had we mentioned the fact that at least seven other independent researchers had considered such a structure. But the future Laureates never stopped to think that such a structure had ever before been considered, when in fact it was actually a pretty common idea.

**Metallofullerenes**

I remember Harry measuring a 3 foot diameter molecular model and telling us there was room for an atom inside. Knowing as much as we did about iron and steel, we all felt that iron was an obvious atom to put inside. The resulting Fe-$C_{60}$ species would be a very interesting new molecule.

Generating the first metal-$C_{60}$ cluster had been my project for at least two weeks. I tried at least 10 different times to coat the graphite with various iron species. Every attempt failed for reasons we all now understand. But at the time it was beginning to look bad for the soccerball hypothesis.

I tried hard for days to obtain $C_{60}Fe$. I wanted it, I wanted to succeed, I was interested, and I felt challenged. But it didn't work. Jim had worked on carbon lanthanum intercalation compounds during his undergraduate research at Baylor and he wanted to try La. It worked. And for reasons I never understood Rick decided to make Jim first author on a paper which was a direct result of my assigned project.[13]

### Fullerenes

The creation of the field of fullerene chemistry was the most important thing we ever did. And the Swedish Academy agreed. They didn't award the Nobel for $C_{60}$. They awarded it for the fullerenes. Discovering $C_{60}$ was almost obvious in hindsight. Multiple independent researchers had studied the structure. But discovering the fullerenes was a stunning shock to the every chemist in the world. While searching for proof of the soccerball structure we rewrote the rules on carbon chemistry. This was another example of finding something we weren't looking for.

As a summer student in 1984 in Rick's lab I had used the reaction nozzle to work on metal cluster reactions.[2-3] I used it again to try prove that $C_{60}$ was special and everything else was highly reactive. I failed at that one too, but this failure turned into a Nobel Prize.

The reaction chemistry studies repeatedly confirmed the low reactivity of the even carbon clusters. All of them were inert, not just $C_{60}$. For at least seven consecutive days in October 1985 I went into Rick's office and presented him clear evidence that all the cages were closed. He refused to believe it. Over and over he refused to accept these results because of his passionate belief that $C_{60}$ was special, the closed $C_{70}$ cage was an anomaly, and all other carbon clusters were open cages with highly reactive dangling bonds.

A few days later he grudgingly accepted the results, then showed me a preprint of a Haymet paper on $C_{60}$ and $C_{120}$.[14] Rick and I together, using the equations in Haymet's paper, predicted 12 pentagon closed cages for any even number of atoms. This was the most important moment in my scientific career. Euler had learned this rule several centuries earlier, but we were the first chemists to apply it to real molecules.

The fullerene paper[15] completed our trio of submissions. 1985 was our annus mirabilis. Over a period of three months we created a new field of chemistry, we transformed the way people thought about carbon, and we established the Smalley lab as the premier physical chemistry laboratory in the world.

We began this work attempting to study the reaction of small carbon chains. We did finally publish the article which was the original goal of the experiment.[16] It is a wonderful article, deeply rooted in astrophysical carbon chemistry. But it is now pretty much just a footnote.

**Photophysics of the Fullerenes**

The crown jewel of my thesis research was studying the decay paths of metastable fullerene cations.[17,18] Most photodissociation events occurred in nanoseconds, or perhaps microseconds. I was able to manipulate the tool to study fullerene cations which were disintegrating in milliseconds. One version of this dissociation path was shown in a panel in a *Scientific American* article on the fullerenes.[19]

**Conclusions**

The fall of 1985 was a magical time for physical chemistry and for the members of the fullerene team. By 1988, at the time of my graduation, most chemists were of the opinion that we were right but that $C_{60}$ was nothing more than an interesting diversion because milligrams of fullerenes were not available. The discovery of the carbon arc synthesis method ignited fullerene research and established our place in history.[20]

## Acknowledgments

The main point of this paper is to document my activities during the discovery of the fullerenes. It could span 100 pages if I tried to document what everyone else did. This paper in no way is meant to imply that others played no role, or unimportant roles. Yuan Liu wrote key portions of the Zeus software. Qingling Zhang contributed as much effort on the fullerene paper[15] as I did, earning her place as first author. Mike Guesic and Mike Morse taught me how to run AP2 and specifically how to use the downstream fast flow cluster reactor. Falk Weiss, Tapani Laaksonen, Mike Alford, and Jerry Elkind were the key people on the ICR, which enabled our photofragmentation studies. Claire Pettiette and Phil Brucat were critical to the fast startup of the AP2 tandem mass spectrometer.

December 2012

## References

1. B L Cousins et al., Argon isotope effect in the microwave spectra of ArHF, *J Phys Chem*, **88**, 5142 (1984).
2. M E Geusic et al., Surface reactions of metal clusters I: the fast flow cluster reactor, *Rev Sci Inst*, **56**, 2123 (1985).
3. M D Morse et al., Photofragmentation processes in metal-ligand complexes: benzene-tungsten and bis-benzene-tungsten, *Chem Phys Lett*, **122**, 289 (1985).
4. E A Rohlfing et al., Production and characterization of supersonic carbon cluster beams, *J Chem Phys*, **81**, 3322 (1984).
5. S C O'Brien et al., Supersonic cluster beams of III–V semiconductors: $Ga_xAs_y$, *J Chem Phys*, **84**, 4074 (1986).
6. J Baggott, *Perfect Symmetry*, Oxford University Press (1996).

7. H W Kroto, J R Heath, S C O'Brien, R F Curl, and R E Smalley, $C_{60}$: Buckminsterfullerene, *Nature*, **318**, 162 (1985).

8. H P Schultz, Topological organic chemistry, polyhedranes and prismanes, *J Org Chem*, **30**(5) 1361 (1965).

9. Z Yoshida and E Osawa, *Aromaticity*, Chemical Monograph Series 22. Kyoto: Kagaku-dojin, p. 174 (1971).

10. D A Bochvar and E G Galpern, *Doklady Phys*, **209**, 610 (1973).

11. J Castells and F Serratosa, Goal, *J Chem Education*, **60**, 941 (1983).

12. R A Davidson, *Theor Chim Acta*, **58**, 193 (1981).

13. J R Heath et al., Lanthanum complexes of spheroidal carbon shells, *J Am Chem Soc*, **107**, 7779 (1985).

14. A D J Haymet, $C_{120}$ and $C_{60}$: Archimedean solids constructed from $sp^2$ hybridized carbon atoms, *Chem Phys Lett*, **122**(5), 421 (1985).

15. Q Zhang et al., Reactivity of large carbon clusters: spheroidal carbon shells and their possible relevance to the formation and morphology of soot, *J Phys Chem*, **90**, 525 (1986).

16. J R Heath et al., The formation of long carbon chain molecules during laser vaporization of graphite, *J Am Chem Soc*, **109**, 359 (1987).

17. S C O'Brien et al., Photophysics of buckminsterfullerene and other carbon cluster ions, *J Chem Phys*, **88**, 220 (1988).

18. F D Weiss et al., Photophysics of metal complexes of spheroidal carbon shells, *J Am Chem Soc*, **110**, 4464 (1988).

19. R F Curl and R E Smalley, Fullerenes, *Scientific American*, **265**, 45 (1991).

20. H Aldersey-Williams, *The Most Beautiful Molecule: The Discovery of the Buckyball*, Wiley (1997).

# A7

# How I conceived soccerball molecule $C_{60}$

## Eiji Ōsawa

*Nanocarbon Research Institute Limited, Asama Research Extension Centre,*
*Faculty of Textile Science and Technology, Shinshu University, 3-15-1 Tokida,*
*Ueda, Nagano 386-8567, Japan*
osawaeiji@aol.com, osawa@nano-carbon.jp

It is great honor to contribute a note to the book commemorating the silver anniversary of the discovery of buckminsterfullerene $C_{60}$.[1] My actual contribution to the discovery was virtually nil: I only thought about $C_{60}$ somewhat earlier.[2,3] However, this small action of conceiving an unknown molecule changed my life after it was shown to really exist. I never expected that my fancy dream would ever come true in the real world. At first I thought I was very, very lucky. However, later I analyzed the incident more carefully and found that it happened as it should have. In this small note I will try to describe the route to my conception.

*$C_{60}$ Buckminsterfullerene: Some Inside Stories*
Edited by Harry Kroto
Copyright © 2015 Pan Stanford Publishing Pte. Ltd.
ISBN 978-981-4463-71-3 (Hardcover), 978-981-4463-72-0 (eBook)
www.panstanford.com

## Encounters in Princeton

First I must mention my background in "conceptual chemistry." When I began my postdoctoral training in the laboratory of Prof. Paul Schleyer at Princeton University in 1967, I met Prof. Kurt Mislow, Paul's closest friend in the same department. In our first conversation, Kurt curtly asked me, "What is your principle?" I had never been asked something like that before and I could answer only vaguely, something like: "To make new compounds and extract novel properties." Then Kurt said, "New compounds are not that important. The most important thing in your work is to create a new concept." I was shocked, as this assertion is the exact opposite of what I had been taught in Japan. Then he told me how he plans his work on stereochemistry with grand perspectives.[4] He was a perfectionist as well. He told me that papers must be perfect not only in English grammar but also in style and logic. All these lessons made a tremendous impact upon a young and naïve man like me.

Naturally, concepts must be original and brewed carefully in your head. Brewing your own concepts touches upon your basic attitude towards science. I never thought I would be good enough to propose new scientific concepts, but Kurt told me to do that. Why? This is a challenging task and demands one to be very positive in thinking. Then, looking around in Frick Chemical Laboratory, I realized everyone was seeking a new concept. In group seminars, weekly departmental seminars, and in all conversations in and out of the laboratory, people were so eager to interpret their experimental results from new angles and propose new ideas. I felt I was in the right place.

The best thing I learned at Princeton from Kurt and Paul, and many other excellent people (Richard Smalley was there as a graduate student, and Peter Stang as a superpostdoc!), is that your new concept must be reasonable and acceptable to others; in other words, you must follow a track that predecessors have explored and proved correct. The next important lesson I learned was that you must be imaginative and open to the discussion of your ideas with the people around you.

**3D Aromaticity**

I returned to Japan in 1969 with a very clear mission in mind, which was to conceive something original, feasible, and interesting. In those days one of the most popular topics in organic chemistry (I am an organic chemist) was non-benzenoid aromatics. Nozoe's 7-membered troponoids and Sondheimer's $4n + 2$ monocyclic conjugated systems were the favorites of many of my colleagues, but it seemed to me that they were simply putting time and energy into the synthesis of *new compounds* designed according to Hückel's $4n + 2$ rule.[3,5] The targets were planar, $\pi$-conjugated 2D molecules. Although the new non-benzenoids were beautiful and exciting, I chose to try to conceive of a new aromatic system which could display much more dramatic effects than 2D delocalization.

As a consequence, instead of working in lab, I read extensively about novel ways of electron delocalization. I soon stumbled upon Muetterties's polyhedral boranes,[6] which seemed wonderful alternatives to the planar aromaticity concept. Then I noticed Lipscomb's pioneering works on simple borane compounds and the work of Roald Hoffmann that he did on boranes using his extended Hückel theory, developed while he was a graduate student of Lipscomb.[3] All this pertained to the aromaticity concept in a broader sense. These studies seemed bad news to me as there seemed to be nothing left for me to contribute.

## Spherical Aromatic Carbon

Disappointed, I returned back from the boranes to my old hydrocarbon playground, but could not find any structure that might exhibit the remarkable feature of 3D aromaticity. Incidentally I saw Prof. Paul Schleyer's short communication on the first 3D aromatic, the dehydroadamantyl dication. This seemed to be a brilliant solution to the problem I was thinking about and Prof. Schleyer reached the solution using adamantane, which he had himself rediscovered and developed. I have already commented on this work before.[2] Anyhow, the territory I was trying to defend, 3D aromaticity, was rapidly shrinking.

It was just about this time that I stepped down from 3D to spherical aromatics and thought of $C_{60}$.[2] I anticipated that spherical aromaticity would be much less remarkable compared to real 3D aromaticity. The similarity of spherical aromaticity in polyhedral boranes was always on my mind, and $C_{60}$ can be viewed as a carbon analogue of the same.[3] Even now, we can readily pick out a few candidates for spherical aromatic carbon species by looking at a collected set of illustrations of known polyhedra and choosing feasible structures of median polyhedra (three-edged vertices with 4–8 polygonal network facets). For me, $C_{60}$ was a compromise. This conclusion is certainly one of the reasons why I did not pursue $C_{60}$ further after 1970.

## Lucky Incident

Even this backyard race was tough, and I am sure I could have never entered the competition if it were not for a few lucky happenings. Most of them have been told, but recently I learned a new fact from Prof. H. Hosoya.[7] FIFA (Fédération Internationale de Football Association) adopted the now popular truncated icosahedron design

**Figure 1** *Left*: One of the old FIFA official soccerball designs made of rectangular leather patches stitched together. *Right*: Since 1970 World Cup Championships use the Archimedean design made up with pentagons and hexagons of artificial leather patches. Images down loaded from internet.

for its official ball in *1970* in place of the old hexahedral patterns (Photo 1). It is highly likely that the sports and toy shops all over the world began selling the footballs of a new design just before the 1970 World Cup Championship in Mexico opened. I saw one such football in my son's hands when I was trying to combine corannulene patterns into a sphere.[2] Without this incident, the idea of $C_{60}$, would have been much delayed.

Another incredible incident occurred after the discovery of $C_{60}$ by Kroto, Heath, O'Brien, Curl, and Smalley became known. Dr. Tadamichi Fukunaga, who at the time was in the Central Research Laboratory of Du Pont Co., Delaware, happened to buy a copy of my book *Aromaticity*, written in Japanese,[3] when he was back home in Japan on holiday, and kindly thought of sending it to Prof. Smalley with partial translation. He asked Prof. W. Herndon from the University Texas at El Paso, who happened to visit Du Pont's Central Research, to carry out this mission, perhaps because both El Paso and Rice are in Texas. Actually they are 1500 km apart, and it had to pass through the hands of Doug Klein, and Nenad Trinajstic of Croatia, before reaching Kroto.

Prof. R. F. Curl has told this story.[8] In this way I became acquainted with Kroto, Smalley, and Curl, three of the nicest people I have ever met. The most remarkable feature of this incident is that I had not met with these people before, including Dr. Fukunaga. They were all kind enough to appreciate my small breakthrough of conceiving the unique truncated icosahedral structure of C$_{60}$.

**Fullerene Fever**

Even after the discovery, I was not competent enough to enter the race to isolate and characterize C$_{60}$,[9] and watched, with a feeling of awe, as people thought and worked on the system during the difficult period from 1985 to 1990. Then C$_{60}$ and carbon nanotubes were isolated in a short space of time, 1990–1991, and then spectacular developments in "fullerene science" ensued. The targets were all forms of the *closed* 5/6 network of carbons, including giant and strained fullerenes, nanotubes, and onion-like multi-shell structures. One good measure for the feverish research activity is the very high numbers of citations[10] of three key papers:

- 10,151 for the C$_{60}$ discovery paper by H. Kroto et al.[1]
- 6,265 for the C$_{60}$ isolation paper by W. Krätchmer et al.[11]
- 27,570 for the discovery of carbon nanotube by S. Iijima.[12] Iijima has three more papers cited more than 1000 times.[13–15]

**Epilogue**

As I mentioned above, C$_{60}$ was a bitter compromise when I crystallized the structure in my head. However, its tremendous success gave me great confidence in my conceptual approach. I found myself able to spin out something interesting and valuable. Although I was not prepared to join the competitive exploration of fullerene science experimentally, I watched the development carefully. Then I saw amazing proof for Prof. Mislow's words: *new compounds are not as important.* The Nobel-winning discoverers of $C_{60}$ changed their interests into carbon nanotubes, and research activities in $C_{60}$ quickly faded in the late 1990s.

In spite of Kurt Mislow's warning, it would be useful to understand why fullerene research has declined. We soon notice that none of the fundamental problems of fullerenes have actually been completely solved. For example, the formation mechanism is not completely understood. Recently we saw breakthroughs in this direction, one by Dunk et al., who have obtained experimental evidence for the "traditional" deterministic approach,[16] and the other by Irle et al. for the nontraditional nondissipative approach.[17] However, final complete answers are still elusive. For this reason, $C_{60}$ is still the only fullerene that can be mass-produced, and the size and chirality of CNTs are uncontrollable. It seems that our present chemical synthesis techniques are not adequate for the creation of such complex carbon networks and some new methodology is needed to create these species. Larry Scott, who also has contributed to this set of personal accounts, has made the first inroads into solving these very difficult technical problems. In this regard, fullerene research may be seen as only having just begun and the apparent decline may only be a transitional moment.

## Acknowledgements

Thanks are due to Profs. Istvan Hargittai, Robert Curl, Harry Kroto, Zdenek Slanina, and (the late) Richard Smalley for constant encouragement of the author since the emergence of $C_{60}$.

## References and Notes

1. H W Kroto, J R Heath, S C O'Brien, R F Curl, and R E Smalley, "C$_{60}$: Buckminsterfullerene," *Nature*, **318**, 162–163 (1985).

2. E Ōsawa, "The evolution of the football structure for the C$_{60}$ molecule: a retrospective," in *The Fullerenes, New Horizons for the Chemistry, Physics and Astrophysics of Carbon*, ed. H W Kroto and D R M Walton, Cambridge University Press, Cambridge, pp. 1–8 (1993).

3. Z Yoshida and E Ōsawa, *Aromaticity* (in Japanese), No. 22 of Chemical Monograph Series, Kagaku Doujin, Kyoto, p. 191 (1971).

4. K Mislow, *Introduction to Stereochemistry*, W A Benjamin, Inc., New York, p. 193 (1966).

5. *Nonbenzenoid Aromatics*, ed. J P Snyder, Academic Press, New York, p. 434 (1971).

6. E L Muetterties and W H Knowth, *Polyhedral Boranes*, Marcel Dekker, Inc., New York, p. 197 (1968).

7. H Hosoya, "Soccer-ball in cosmos" (in Japanese), *Proceeding of the 76th National Science Education Conference*, **76**(2), 17–23 (2004).

8. R F Curl, R E Smalley, H W Kroto, S O'Brien, and J R Heath, "How the news that we were not the first to conceive of soccer ball C$_{60}$ got to us," *J Comput Graph Model*, **19**, 185–186 (2001).

9. H Aldersey-Williams, *The Most Beautiful Molecule: An Adventure in Chemistry*, Aurum Press, London, p. 340 (1995).

10. Google Scholar, as of February 5, 2013.

11. W Krätchmer, L D Lamb, K Fostiropoulos, and D R Huffman, "C$_{60}$: a new form of carbon," *Nature*, **347**, 354–358 (1990).

12. S Iijima, "Helical microtubules of graphitic carbon," *Nature* 1991, *354*, 56–58.

13. S Iijima and T Ichihashi, "Single-shell carbon nanotubes of 1-nm diameter," *Nature*, 363, 603–605 (1993).

14. K Hata, D N Futaba, K Mizuno, T Namai, M Yumura, and S Iijima, "Water-assisted highly efficient synthesis of impurity-free single-walled carbon nanotubes," *Science*, **306**(5700), 1362–1364 (2004).

15. S Bae, H Kim, Y Lee, et al., "Roll-to-roll production of 30-inch graphene films for transport electrodes," *Nat Nanotechnol*, **5**, 574–578 (2010).

16. P W Dunk, N K Kaiser, C Hendrickson, J P Quinn, C P Ewels, Y Nakanishi, Y Sasaki, H Shinohara, A G Marshall, and H W Kroto, "Closed network growth of fullerenes," *Nat Commun*, **3** (2012), DOI:10.1038/ncomms1853.

17. S Irle, G Zheng, Z Wang, and K Morokuma, "The $C_{60}$ formation puzzle 'solved': QM/MD simulations reveal the shrinking hot giant road of the dynamic fullerene self-assembly mechanism," *J Phys Chem B*, **110**, 14531–14545 (2006).

# Partial translation of the book *Aromaticity* (in Japanese)

## Zen-ichi Yoshida and Eiji Ōsawa*

*Nanocarbon Research Institute Limited, Asama Research Extension Centre, Faculty of Textile Science and Technology, Shinshu University, 3-15-1 Tokida, Ueda, Nagano 386-8567, Japan*

osawaeiji@aol.com, osawa@nano-carbon.jp

### 5.6.2 The Possibility of Superaromatic Hydrocarbons

The term *superaromaticity* is temporarily defined here as a significant lowering of the potential energy of a molecule as the result of 3D delocalization of electrons in high-energy bonding molecular orbitals. First of all we consider whether there is any possibility of realizing such a phenomenon in familiar hydrocarbon frameworks.

To start with, let us take a look at the back cover of *Cram's Textbook of Organic Chemistry*, where several "dream molecules" are depicted. Among these, truncated tetrahedrane $C_{12}H_{12}$ (**147**) appears to be capable of its cyclopropane bonds with high p character interacting with each other over the molecular surface. Geometrically, **147** is obtained by truncating four vertices of the hydrocarbon tetrahedrane (**148**).[a] As it is less strained than the

---

* Kagaku Doujin, Kyoto, 1971, pp. 174–178.

elusive species **148** and with the possibility of stabilization by 3D aromaticity, this molecule is currently the target of synthesis by several research groups.[b]

Let us turn to some more likely cases for superaromaticity, involving $p_z$ orbitals perpendicular to a spherical molecular surface rather than the overlap involving σ orbitals. If we are to achieve superaromaticity by delocalization of π electrons over a positively curved molecular surface, such a molecule has to be large enough to overcome the disadvantage from the reduced overlap of nonparallel $p_z$ orbitals. Following the same strategy of truncating a Platonic solid to produce a spherical structure, we came across a potential candidate structure, the icosahedron (**149**). Truncation of a vertex produces a regular pentagon. Thus, truncation of all 12 vertices gives a beautiful 32-faced solid (**150**), whose geometric name is the truncated icosahedron.[c] The original 20 triangular faces in **149** are replaced by 20 hexagons, which surround and isolate the 12 pentagons generated at the positions of the original vertices. The angles at the vertices in the component polygon do not deviate too much from planar, and all of the 90 edges seem almost equal in length as judged by the model. Thus, the virtual $C_{60}$ molecule[d] which can be derived by replacing 60 vertices of **150** with $sp^2$-hybridized carbon atoms does not seem totally unrealistic.

Interestingly enough, Barth and Lawton[30] recently synthesized a molecule which corresponds exactly to a part of the surface design of **150**: dibenzo[ghi,mno]fluoranthene or corannulene (**151**), which forms almost colorless prismatic crystals melting at 268 to 269°C. According to single-crystal X-ray analysis, 151 is a bowl-shaped structure, and hence does indeed correspond to a partial structure of **150** (**Figs. 5.6 and 5.7**). However, before X-ray analysis was carried out, Barth and Lawton could not exclude a planar structure for the following reasons: An alternant hydrocarbon containing an odd-membered ring like 151 cannot have a uniform π-electron density distribution in its ground state.[e] One of such polar structures is the double Hückel aromatic structure (**151b,c**) consisting of peripheral

14π- and central 6π-electron systems and the cyclic stabilization contribution as shown in **151c** should be a maximum when the whole molecule is planar.[f]

Nevertheless, the planar structure of **151** suffers from large angle strain. If we assume that all C–C bonds are 1.40 Å in length, and the central ring is a regular pentagon for the planar corannulene, angle $\theta$ shown in **151a** would have a formidable value of 144°! The $\theta$ value may recover the standard value of 120° if the five bonds extending from pentagon deformed out of plane by 38° as illustrated in **152**. In such a deformed structure, however, the neighboring benzene rings cannot be coplanar. At this point we should note that the corannulene molecule provides an ideal framework for addressing an interesting question: How much 2D aromaticity would be lost if the molecule were distorted to take on a nonplanar 3D structure? The answer to this question is the key to the superaromaticity of $C_{60}$.

Let us study the observed structure of corannulene more carefully (**Fig. 5.7**). The dihedral angle between a pentagon and a hexagon is 26.8° in average. Therefore considerable strain remains in the structure. The observed $\theta$ angle between six-membered rings is 130.9°. Namely the corannulene is a shallow bowl somewhere between a planar (**151**) and a deep bowl (**152**). Proton NMR spectrum ($\tau$ 2.19) indicates that considerable aromaticity remains. Corannulene forms weakly-colored charge-transfer complexes with picric acid and trinitrobenzene.[30] It is still not clear how much of the resonance energy of the planar form is lost on taking on the shallow-bowl structure. According to SCF-LCAO-MO calculations by Gleicher,[31] even the deep-bowl structure (**152**) retains more than 90% of the aromaticity of the planar form.

### Notes and References

a. **147**: heptacyclo$[5.5.0.0^{2,12}.0^{3,5}.0^{4,10}.0^{6,8}.0^{9,11}]$dodecane, $T_d$ point group.

b. (a) R B Woodward and R Hoffmann, *The Conservation of Orbital Symmetry*, Academic Press, New York (1970), p. 106. (b) E Baldwin and F El-Barkawi, 159th ACS National Meeting, Petroleum Chemistry Division, Abs No. 23, Houston, TX, February (1970).

c. This pattern appears on the surface of all of the official soccer balls designs by FIFA (Fédération Internationale de Football Association) for use in the World Cup Championship since 1970.

d. (Note added in translation) In the book, "$C_{60}H_{60}$" appears here by typographical error.

e. See Chapter 1.2.3 of the book.

f. Kekulè structure of corannulene formally corresponds to coronene (**153**, a classical planar aromatic molecule) minus one benzene ring.

(30) W E Barth and R G Lawton, *J Am Chem Soc*, **93**, 1730–1745 (1971).

(31) G J Gleicher, *Tetrahedron*, **23**, 4257–4263 (1967).

# The first stepwise chemical synthesis of C$_{60}$

## Lawrence T. Scott

*Merkert Chemistry Center, Boston College, Chestnut Hill, MA 02467-3860, USA*

lawrence.scott@bc.edu

By the summer of 1989, much of the initial euphoria surrounding the discovery of C$_{60}$ had waned. The sudden burst of scientific publications had dwindled to a trickle, down to one paper every month or so, and none described new experimental work. Nearly four years had passed since this spectacular new allotrope of carbon had been identified as a totally unexpected product of graphite laser ablation, but nobody could get their hands on it. We all knew it was there. The experimental evidence was incontrovertible. It had been seen in the gas phase by mass spectrometry, and its electron affinity, UV, and photoelectron spectra had been determined. Did it really have the soccer ball shape? We could not prove it, but what else could it be? Why could

*C$_{60}$ Buckminsterfullerene: Some Inside Stories*
Edited by Harry Kroto
Copyright © 2015 Pan Stanford Publishing Pte. Ltd.
ISBN 978-981-4463-71-3 (Hardcover), 978-981-4463-72-0 (eBook)
www.panstanford.com

$C_{60}$ not be found when the apparatus used to generate it was opened and its contents were examined? One obvious conclusion was that $C_{60}$ must be too reactive to exist in monomeric form in the condensed state. Surely the strain energy resulting from the enforced pyramidalization of all those trigonal carbon atoms would be relieved significantly by joining together two or more fullerenes through covalent bonds, thereby transforming strained, unhindered trigonal carbon atoms into unstrained, tetrahedral carbon atoms. In short, maybe $C_{60}$ simply polymerized in the solid state. Many scientists accepted this perfectly reasonably conclusion and wrote off $C_{60}$ as a marvelous but ephemeral gas phase species that would never survive captivity.[1]

I was not one of those skeptics. Pyramidalized trigonal carbon atoms were already well known in such exotic hydrocarbons as Virgil Boekelheide's "superphane" (**1**) and Fred Greene's tetradehydrodianthracene (**2**), both of which had been synthesized, isolated, and characterized as stable monomeric species under ordinary laboratory conditions in the 1970s. As a hydrocarbon chemist by birth,[2] I knew this literature well. It seemed reasonable to me that $C_{60}$, in which every pyramidalized carbon atom belongs to a benzene ring, could be stable, too. No doubt, I was influenced by my senior colleagues at UCLA: Don Cram, whom many regard as the father of cyclophane chemistry, and Orville Chapman, who had a model of $C_{60}$ on his desk, well before 1985, plus graduate students in his laboratory

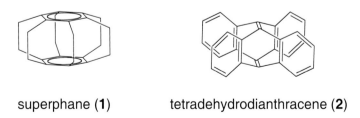

superphane (**1**)        tetradehydrodianthracene (**2**)

**Figure 1** Hydrocarbons from the 1970s containing pyramidalized trigonal carbon atoms.

pursuing projects designed to synthesize $C_{60}$ by stepwise chemical methods.[3]

Then I saw Harry Kroto for the first time. He was one of the plenary lecturers at the 6th International Symposium on Novel Aromatic Compounds in Osaka, Japan, in August 1989. He talked about "$C_{60}$, Fullerenes, Giant Fullerenes, and Soot" and recounted some of the early experiments involving $C_{60}$.[4] I was mesmerized! When he mentioned in passing that other scientists had detected $C_{60}$ in fuel-rich flames, I nearly fell out of my chair. Flames? Combustion? How could burning a hydrocarbon like benzene in a fuel-rich (i.e., oxygen-deprived) atmosphere make $C_{60}$? The chemistry going on there must be totally different from that in the original Kroto/Smalley experiments involving lasers and graphite. In the combustion of benzene, it seemed to me, benzene molecules would have to dimerize to make biphenyl and lose two hydrogen atoms. As more benzene rings became attached to the nascent soot particle, intramolecular bonds would have to form new rings, liberating more hydrogen atoms. Such cyclodehydrogenation reactions would have to form both 5- and 6-membered rings, thereby imposing curvature, and eventually all the hydrogens would have to be lost.

Despite the vast difference between these two routes to $C_{60}$, it struck me that they shared two features in common. Both involved chemical transformations under solvent-free conditions in the gas phase, and both involved temperatures far above those used in everyday chemistry. Such conditions are totally foreign to traditional synthetic organic chemists, but they were quite familiar to me. Since the middle of 1970s, my laboratory had been studying the high-temperature gas phase chemistry of aromatic hydrocarbons using a venerable technique known as flash vacuum pyrolysis (FVP).[5] In these experiments, organic compounds are heated to very high temperatures for fractions of a second by passing them rapidly through a hot quartz tube at low pressures in the gas phase and collecting the products in a cold trap. Benzene has long been known to dimerize to biphenyl at high temperatures in the gas phase **(Scheme 1)**.[6] We had also encountered several simple *intramolecular* aryl–aryl oxidative coupling reactions that gave either 5- or 6-membered rings under FVP conditions **(Scheme 1)**.

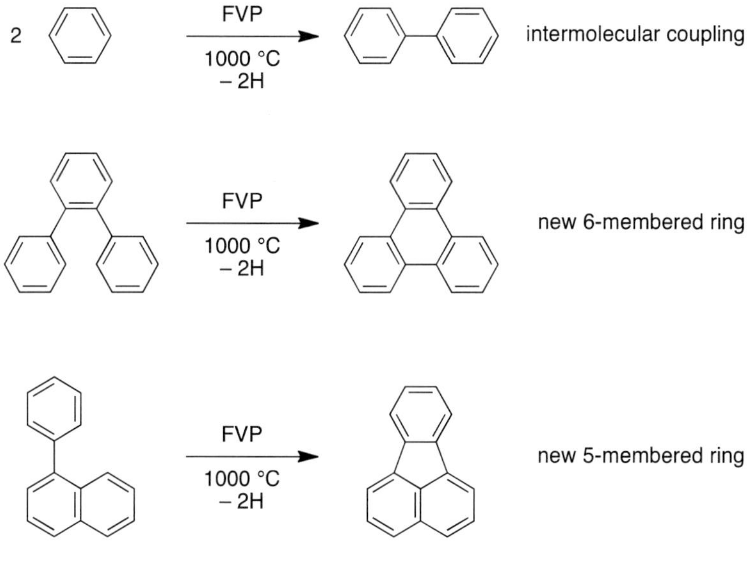

**Scheme 1** Intermolecular and intramolecular coupling reactions of benzene rings at high temperatures in the gas phase.

Suddenly, the formation of C$_{60}$ in flames did not seem so unreasonable after all! The intermolecular and intramolecular C–C bond-forming reactions were all well precedented **(Scheme 1)**, and the high temperatures could provide enough energy to distort the growing polycyclic aromatic hydrocarbons (PAHs) away from their preferred planar geometries. Temporarily bending the PAHs could bring remote portions of the molecules close enough together to form additional bonds, which would irreversibly enforce geodesic curvature. At such high temperatures, the entropic gain (*T*$\Delta$*S*) of losing two more hydrogens as each new C–C bond formed could outweigh the enthalpic cost ($\Delta$*H*) of gradually incorporating additional strain into the growing PAH on its inexorable cascade *down* ($\Delta$*G* < 0) to C$_{60}$.

It was there, in Osaka, listening to Harry Kroto, when it first occurred to me that FVP might be the perfect tool for synthesizing C$_{60}$ (and other fullerenes) by rational design in the laboratory. In principle, we should be able to build a planar 60-carbon PAH with all the atoms connected in a pattern that mapped perfectly onto C$_{60}$ and then "stitch it up" to make C$_{60}$ by thermal cyclodehydrogenations under FVP conditions. With 150 years worth of accumulated knowledge at our disposal about how to build planar PAHs, we had no doubts about our ability to assemble a suitable 60-carbon precursor for the final FVP step. The only real question, then, was whether or not FVP could be used to stitch it up to make C$_{60}$. It seemed entirely reasonable to us, but would it work?

Through the entire night on the long flight home from Osaka, I feverishly scribbled plans to disconnect enough C–C bonds in C$_{60}$ to come up with a planar PAH that might serve as a synthetic precursor. My oldest daughter, Jenny, accompanied me on this trip to Japan, and she woke up every few hours, looking amused to find me still filling page after page of paper with my drawings. I was obsessed!

Finally, somewhere over the Pacific Ocean, I came to the realization that the entire plan hinged on an untested hypothesis. Could FVP really be used to transform a planar PAH into a geodesic polyarene? There was zero precedent for such a transformation. We needed to demonstrate that first. If the concept could be validated, then all the pieces would be in place to synthesize C$_{60}$ by rational design in the laboratory.

We chose corannulene (**4**), the smallest subunit of C$_{60}$ that retains geodesic curvature, as the ideal test molecule. Corannulene was, in fact, already a known compound long before the discovery of C$_{60}$.[7] Richard Lawton, an assistant professor at the University of Michigan, dreamed up the idea of corannulene and then synthesized the compound in the laboratory by a beautiful (but long, 18 steps) sequential annulation strategy. It took us some time to design and synthesize the right FVP precursor to test our hypothesis, but ultimately, we succeeded. FVP of 7,10-diethynylfluoranthene (**3**) produced corannulene (**Scheme 2**).[8] A geodesic polyarene had been synthesized from a completely planar PAH using FVP to temporarily bend the molecule and bring remote atoms close enough together to form new C—C bonds. The strategy worked, and we were off on our journey.

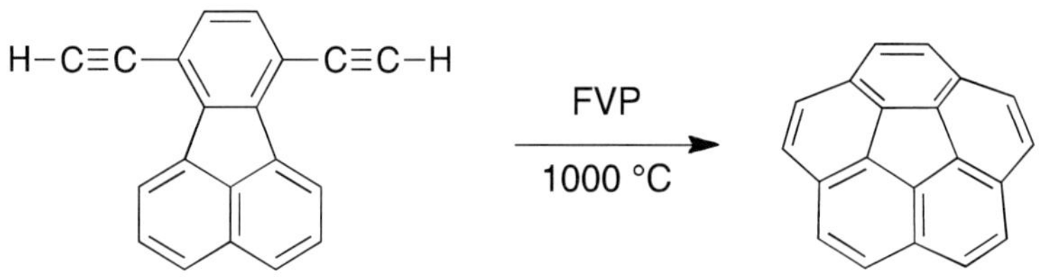

H−C≡C— —C≡C−H

FVP
1000 °C

7,10-diethynylfluoranthene (**3**)          corannulene (**4**)

**Scheme 2** Flash vacuum pyrolysis (FVP) synthesis of corannulene (**4**). The first conversion of a planar polyarene to a geodesic polyarene.

The corannulene synthesis in **Scheme 2** is actually a thermal isomerization reaction. No hydrogen atoms are lost. This isomerization does not require the entropic benefit of losing hydrogen atoms, because corannulene is actually more stable thermodynamically than the diyne precursor **3**. The enthalpic gain from forming two new, strong sp$^2$–sp$^2$ C—C bonds and two new aromatic benzene rings at the expense of two weak alkyne $\pi$ bonds is more than enough to pay for the strain introduced. We eventually shortened our FVP-based corannulene synthesis to just three steps.[9]

Emboldened by this triumph, we went on to demonstrate that FVP could be used to stitch up planar PAHs also by thermal cyclodehydrogenations **(Scheme 3)**.[10] In one step from a PAH that had been known since 1883,[11] we could build 60% of the $C_{60}$ ball by FVP. Further refinements of the FVP strategy eventually gave birth to dozens of geodesic polyarenes,[12] and a new branch of organic chemistry opened up. We almost forgot about synthesizing $C_{60}$!

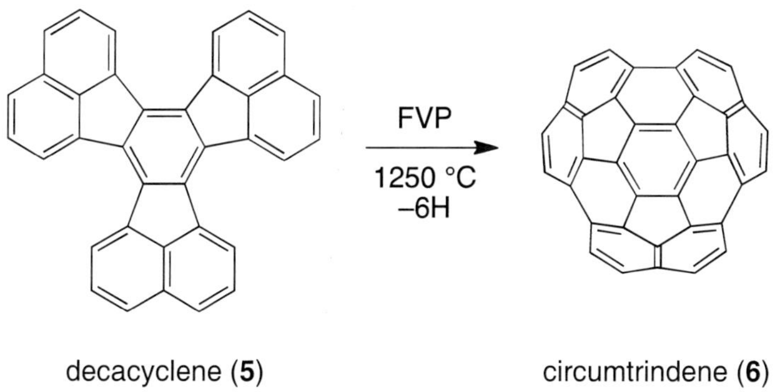

decacyclene **(5)**          circumtrindene **(6)**

**Scheme 3** Circumtrindene, 60% of the $C_{60}$ ball (**6**), synthesized by gas phase thermal cyclodehydrogenation of decacyclene (**5**).

It was Meg Boorum, a PhD student in my research group, who finally set out to synthesize a 60-carbon PAH that could be stitched up to make $C_{60}$. The details of her efforts have already been published,[13] but the bottom line is that she succeeded. By stepwise chemical methods, Meg synthesized the $C_{60}H_{27}Cl_3$ compound **7**, and FVP of **7** produced $C_{60}$ in quantities that could be isolated and characterized spectroscopically **(Scheme 4)**. Unlike all previous syntheses of $C_{60}$, ours gave $C_{60}$ as the only fullerene formed. We saw no trace of $C_{70}$ or of any higher fullerenes. The arms of the precursor are simply stitched together, with loss of all the hydrogen and chlorine atoms on the edges. Our earlier experiences building smaller geodesic polyarenes had taught us valuable lessons about the importance of incorporating bromine or chlorine atoms at strategically chosen locations in our FVP precursors,[12] but such details are beyond the scope of this account.

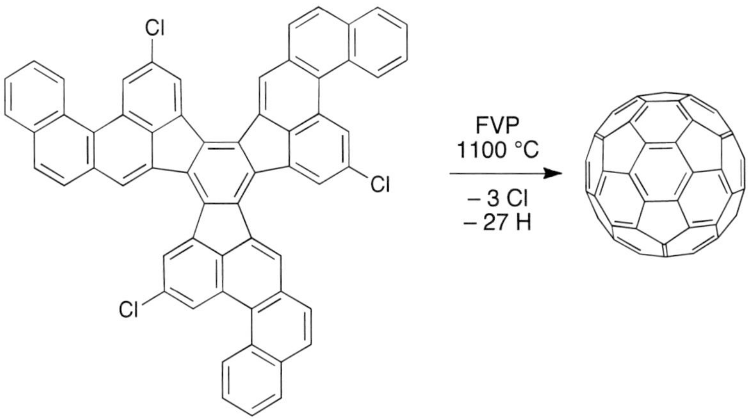

**Scheme 4** Stitching up a 60-carbon precursor (**7**) by flash vacuum pyrolysis to complete the first stepwise chemical synthesis of C$_{60}$.

In the last decade, my research group has turned its attention to the problem of developing methods for the chemical synthesis of single-index $(n,m)$ carbon nanotubes from the ground up.[14] We are no longer in the fullerene synthesis business. I have been pleased, however, to see that others have picked up our strategy for synthesizing C$_{60}$ and have begun applying it to the isomer-specific synthesis of higher fullerenes.[15] Advances have also been made on the isomer-specific synthesis of fullerenes and heterofullerenes using surface-catalyzed cyclodehydrogenation (SCCDH) reactions as a potentially scalable condensed phase alternative to FVP.[16] Organic chemists have an obligation to the rest of the scientific community to figure out ways to "tailor-make" whatever fullerenes or carbon nanotubes anyone might need. We are a long way short of that goal, but the obstacles are not insurmountable. Future generations of organic chemists will surely solve these challenging problems.

## References

1. We now know that $C_{60}$ does not spontaneously polymerize, but this is exactly what happens to smaller fullerenes that violate the isolated pentagon rule, e.g., $C_{36}$: M Menon and E Richter, "Structure and stability of solid $C_{36}$," *Phys Rev B Condens Matter Mater Phys*, **60**(19), 13322–13324 (1999).

2. (a) L T Scott, "Synthesis and photochemical conversion to bullvalene of bicyclo[4.2.2]deca-2,4,7,9-tetraene," undergraduate senior thesis, Princeton University, Princeton, NJ (1966). (b) L T Scott, "Synthetic approaches to several interesting $(CH)_{12}$ hydrocarbons," PhD dissertation, Harvard University, Cambridge, MA (1970).

3. (a) R H Jacobson, "Searching for the soccer ball: An unsuccessful synthesis of corannulene," PhD dissertation, University of California, Los Angeles (1986). (b) Y Xiong, "Part A. A new deep-UV photoresist based on photochemistry of 2-aryl-4,5-benzotropone. Part B. Synthesis of [1.1.1]paracyclophane and other strained aromatic compounds, in search for the soccer ball $C_{60}$," PhD dissertation, University of California, Los Angeles (1987). (c) D Loguercio, Jr, "Studies toward a convergent synthesis of $C_{60}$," PhD dissertation, University of California, Los Angeles (1988). (d) D Shen, "Approaches to soccerene ($I_h$ $C_{60}$) and other carbon spheres," PhD dissertation, University of California, Los Angeles (1990).

4. H Kroto, "$C_{60}$, fullerenes, giant fullerenes, and soot," *Pure Appl Chem*, **62**(3), 407–415 (1990).

5. L T Scott, "Thermal rearrangements of aromatic compounds," *Acc Chem Res*, **15**(2), 52–58 (1982).

6. E Clar, *Polycyclic Hydrocarbons*, Academic Press, New York, vol 2, p 3 (1964), and references cited therein.

7. (a) W E Barth and R G Lawton, "Dibenzo[*ghi,mno*]fluoranthene," *J Am Chem Soc*, **88**(2), 380–381 (1966). (b) R G Lawton and W E Barth, "Synthesis of corannulene," *J Am Chem Soc*, **93**(7), 1730–1745 (1971).

8. L T Scott, M M Hashemi, D T Meyer, and H B Warren, "Corannulene. A convenient new synthesis," *J Am Chem Soc*, **113**(18), 7082–7084 (1991).

9. L T Scott, P-C Cheng, M M Hashemi, M S Bratcher, D T Meyer, and H B Warren, "Corannulene. A three-step synthesis," *J Am Chem Soc*, **119**(45), 10963–10968 (1997).

10. L T Scott, M S Bratcher, and S Hagen, "Synthesis and characterization of a $C_{36}H_{12}$ fullerene subunit," *J Am Chem Soc*, **118**(36), 8743–8744 (1996).

11. P Rehländer, PhD dissertation, University of Berlin, Germany (1883), cited by P Rehländer, *Chem Ber*, **36**, 1583–1587 (1903).

12. V M Tsefrikas, and L T Scott, "Geodesic polyarenes by flash vacuum pyrolysis," *Chem Rev*, **106**(12), 4868–4884 (2006).

13. (a) M M Boorum, Y V Vasil'ev, T Drewello, and L T Scott, "Groundwork for a rational synthesis of $C_{60}$: cyclodehydrogenation of a $C_{60}H_{30}$ polyarene," *Science*, **294**(5543), 828–831 (2001). (b) L T Scott, M M Boorum, B J McMahon, S Hagen, J Mack, J Blank, H Wegner, A de Meijere, "A rational chemical synthesis of $C_{60}$," *Science*, **295**(5559), 1500–1503 (2002). (c) L T Scott, "Methods for the chemical synthesis of fullerenes," *Angew Chem, Int Ed*, **43**(38), 4994–5007 (2004).

14. (a) E H Fort, P M Donovan, and L T Scott, "Diels-Alder reactivity of polycyclic aromatic hydrocarbon bay regions: Implications for metal-free growth of single-chirality carbon nanotubes," *J Am Chem Soc*, **131**(44), 16006–16007 (2009). (b) E H Fort and L T Scott, "One-step conversion of aromatic hydrocarbon bay regions into unsubstituted benzene rings: a reagent for the low-temperature, metal-free growth of single-chirality carbon nanotubes," *Angew Chem, Int Ed*, **49**(37), 6626–6628 (2010). (c) L T Scott, E A Jackson, Q Zhang, B D Steinberg, M Bancu, and B Li, "A short, rigid, structurally pure carbon nanotube by stepwise chemical synthesis," *J Am Chem Soc*, **134**(1), 107–110 (2012).

15. (a) K Y Amsharov and M Jansen, "A $C_{78}$ fullerene precursor: toward the direct synthesis of higher fullerenes," *J Org Chem*, **73**(7), 2931–2934 (2008). (b) K Amsharov and M Jansen, "Synthesis of a higher fullerene precursor—an 'unrolled' $C_{84}$ fullerene," *Chem Commun*, (19), 2691–2693 (2009).

16. (a) G Otero, G Biddau, C Sanchez-Sanchez, R Caillard, M F Lopez, C Rogero, F J Palomares, N Cabello, M A Basanta, J Ortega, J Mendez, A M Echavarren, R Perez, B Gomez-Lor, and J A Martin-Gago, "Fullerenes from aromatic precursors by surface-catalysed cyclodehydrogenation," *Nature*, **454**(7206), 865–868 (2008). (b) K Amsharov, N Abdurakhmanova, S Stepanow, S Rauschenbach, M Jansen, and K Kern, "Towards the isomer-specific synthesis of higher fullerenes and buckybowls by the surface-catalyzed cyclodehydrogenation of aromatic precursors," *Angew Chem, Int Ed*, **49**(49), 9392–9396 (2010).

# C$_{60}$, Arizona, Don Huffman, and other stories

## Lowell D. Lamb

*Broadcom Corporation, San Jose, California, USA*
lamb@broadcom.com, lowelldlamb@gmail.com

## 1985

In the summer of 1985 I left a job on Wall Street and moved with my wife to Tucson, where I enrolled in the University of Arizona as a physics graduate student. I had chosen Arizona because of its strong history in astronomy and hoped to find a research opportunity in that field or a related one. My memories of the first few months in Tucson are a blur of adjusting to the summer heat (we arrived in July), entrance exams, and settling into the routine of coursework.

*C$_{60}$ Buckminsterfullerene: Some Inside Stories*
Edited by Harry Kroto
Copyright © 2015 Pan Stanford Publishing Pte. Ltd.
ISBN 978-981-4463-71-3 (Hardcover), 978-981-4463-72-0 (eBook)
www.panstanford.com

In addition to taking standard graduate courses that first semester, I signed up for a specialty course in light scattering taught by Don Huffman. Don had co-authored the standard reference in the field [1] and had long pursued an experimental program focused on the optical properties of small particles of astrophysical relevance. Don was particularly interested in carbon particles, which are widely believed to be the carriers of the 2200 Å extinction feature commonly seen in spectra of the diffuse interstellar medium [2], and this topic furnished many examples during class discussion. Toward the end of the semester I saw a newspaper article on "small particles of carbon" that I thought might interest Don **(Fig. 1)** [3], so I cut it out and brought it to class.

**Figure 1** 1985 *New York Times* article reporting the Sussex–Rice discovery of $C_{60}$.

I had guessed correctly: Don liked the article. It announced, of course, the Sussex–Rice $C_{60}$ paper in *Nature* [4], and in addition to introducing the name *buckminsterfullerene* and describing the experiment, the article included a tantalizing statement that the researchers "surmised that it may be a common form of carbon in interstellar space." I remember distinctly that after reading the article intently, Don sat quietly for a few moments, then looked up and said, "I'll bet that's what we were making in Heidelberg."

## 1986–1988

Don had spent the 1982–1983 academic year on sabbatical in Wolfgang Krätschmer's lab at the Max Planck Institute for Nuclear Physics in Heidelberg, and one project that had received a good deal of their attention was an attempt to precisely match laboratory measurements of "soot" samples to the 2200 Å interstellar-medium feature. The soot in question was produced not by combustion but by the evaporation of carbon electrodes in a helium atmosphere. Different combinations of electrode size, helium pressure, and electric current yielded a variety of absorption-peak widths and positions, and in the course of trying to match the astrophysical data Wolfgang and Don found a set of conditions that produced the now-famous "Kamel" spectrum. Instead of having the expected single peak typical of small-particle extinction, Kamel soot has three distinct bands or "humps" superimposed on the small-particle peak. Don remembered that he and Wolfgang had debated, without resolution, whether the Kamel spectrum was due to "junk," meaning contamination, or to a "new form of carbon." They also obtained a Raman spectrum of Kamel soot, which showed a number of mysterious peaks superimposed on the graphitic carbon background, but still were unable to explain their results and the problem was shelved when Don returned to Tucson.

During most of the 1985–1986 academic year I was busy with coursework and exams, and other than the light-scattering course I spent very little time on carbon particles or fullerenes. It was only after I cleared these preliminary hurdles and joined Don's group late in 1986 that I was able to give any attention to research. By this point I had a better understanding of Don's earlier, cryptic reference to Heidelberg and I undertook the task of looking for $C_{60}$ in Kamel soot. The way forward seemed simple enough: I needed to build an apparatus that (a) produced soot, (b) measured the UV-visible spectrum during production, thus allowing the identification of Kamel-soot conditions, and (c) provided simultaneous mass spectrometry, thus allowing the detection of $C_{60}$. How hard could that be? I sketched the proposed instrument on the second page of my new notebook **(Fig. 2)** and began a review of the $C_{60}$ literature, which by this time had grown to scores of papers.

It didn't take long to figure out that I had seriously underestimated the complexity of the project and that building the envisioned instrument was going to be very hard (i.e., impossible). For example, one of the key elements was a high-speed valve designed to admit pulses of gas from the smoke chamber into the mass-spectrometry chamber. Reading the papers from Rick Smalley's group that described their molecular beam machine revealed that the corresponding valve in the Rice apparatus represented the culmination of five years of continuous refinement. I, on the other hand, had planned to use a Volkswagen fuel-injection valve. I set the all-in-one apparatus aside and instead focused on recovering the technique for producing Kamel soot. This seemingly simple chore turned into a

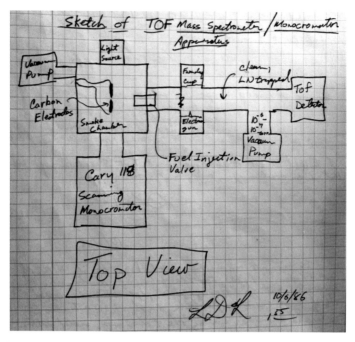

**Figure 2** Initial plan for detecting $C_{60}$ in Kamel soot (from my notebook).

20-month adventure involving a leaky vacuum chamber, a non-functioning spectrometer, a mis-calibrated pressure gauge (I dropped it and didn't tell anyone), and other character-building exercises. By the middle of 1988 I finally had it all working. **Figure 3**, which shows the UV-visible spectrum of one of our first Tucson Kamel soot samples, clearly exhibits three distinct bumps on top of the broad, small-particle peak.

Although it was gratifying to finally master the art of making Kamel soot, in reality I was no closer to understanding this odd material than Don and Wolfgang had been in 1982, and in no way could we connect Kamel soot with $C_{60}$. So I talked with Don in the summer of 1988 and told him that while working on $C_{60}$ was fun, what I really needed was a project that would lead to graduation.

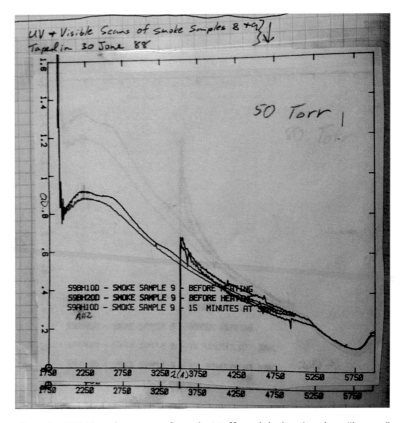

**Figure 3** 1988 Kamel spectrum from the Huffman lab showing three "humps" at 2200, 2700, and 3400 Å (top curve, from my notebook).

He agreed, obtained funding for a light-scattering project, and in the fall of 1988 I set the carbon work aside and began investigating a scattering problem that ultimately grew into the subject of my Ph.D. dissertation.

## 1989

While I was struggling to produce Kamel soot in Tucson, Wolfgang had taken on a new graduate student, Kostas Fostiropoulos, and charged him with the job of learning how to make Kamel soot. Kostas was able to do this with less trouble than I had had, and in 1989 he managed to obtain an infrared absorption spectrum of the soot, which showed four distinct bands superimposed on the graphitic background.

The fact that there were exactly four IR bands was an important clue. Don and Wolfgang had been communicating regularly since the publication of the Sussex–Rice paper and all of us had followed the ballooning buckyball literature. One of the most remarkable theoretical predictions was that $C_{60}$, because of its extraordinary degree of symmetry, would have only four infrared absorption features [5]. Kostas's detection of the four bands was very encouraging, but Don and Wolfgang were not ready to declare victory. First, there was no way to rule out the possibility that the bands were due to some form of contamination, with vacuum-pump oil being suspect number one. Second, the cold-fusion mania was in full swing in 1989 and furnished a clear example of what lay in store for anyone making an important claim without solid evidence. I remember Don saying several times that he didn't want to face the question of whether his $C_{60}$-Kamel soot story was "just some more cold fusion." In the end, Wolfgang presented the experimental results in a paper with the cautious title "Search for the UV and IR spectra of $C_{60}$ in laboratory-produced carbon dust" at a summer cosmochemistry workshop on the Italian island of Capri [6]. Feelings were decidedly mixed: On the one hand, being wrong might prove more than a little embarrassing ("Physicists Discover Pump Oil"). On the other, if Kamel soot really did contain $C_{60}$, there was a chance that the store had just been given away. I'm pretty sure that Don and Wolfgang half hoped that the Capri paper would attract little, if any, attention, at least until we had more time to figure out what exactly was in the Kamel samples.

I also remember that we received a very important visitor that year. Harry Kroto came to give a lecture in the Chemistry Department and what I remember most clearly was his discussion of the larger fullerenes. At one point in his talk he pulled a model of a $C_{540}$ out of a bag and threw it into the audience like a beach ball. This impressed me because I had been so focused on $C_{60}$ and hadn't thought much about anything larger. After the talk Don and I chatted with Harry. Don mentioned that we might have made $C_{60}$ in some of our soot samples, but he couldn't prove it yet. Harry said that it was possible and then the conversation moved on.

## 1990

In the winter of 1989–1990, Kostas followed up his first breakthrough, namely the detection of the four IR bands, with a second and more profound result. Wolfgang had obtained some 99% pure $^{13}C$ powder and Kostas managed to form the powder into rods stable enough to serve as electrodes. I never heard how he did this without using a $^{12}C$-rich binding agent, but I doubt it was easy. Kostas then made two batches of Kamel soot, one with the $^{13}C$ electrodes and the other with natural-abundance electrodes (99% $^{12}C$), and measured their absorption spectra. While the UV-visible spectra of the two samples were virtually identical, the four IR bands in the $^{13}C$ soot had shifted in wavelength from the corresponding $^{12}C$ peaks by a constant ratio. This ratio was almost exactly equal to the square root of the ratio of the $^{12}C$ and $^{13}C$ masses, which indicated that the carrier of the four bands was a pure-carbon molecule. Satisfied that pump oil or any other form of hydrocarbon contamination was ruled out as an explanation of the four IR bands, Don, Wolfgang, and Kostas submitted their results to *Chemical Physics Letters* on the first of May [7], this time with the stronger claim that the isotopic-shift data "supported the idea" that $C_{60}$

was responsible for the four bands. Although Don and Wolfgang would have preferred to sit on the results a bit longer, news that "Harry Kroto is working on your soot" had reached us. It was clear he had taken the Capri paper seriously and that the value of unpublished data might go to zero at any time.

During the month of May those of us in Tucson were preoccupied with the end of the school year. As soon as exams were over, Don was off to Europe to dig out historical measurements of lunar Earthshine which, with proper interpretation, constitute the only existing multi-decadal record of Northern Hemisphere cloud cover, which in turn is an important component of long-term climate models.[1] I was deeply involved in my light-scattering project, and the Tucson summer heat had arrived.

Wolfgang had not been idle. Shortly after posting the *Chemical Physics Letters* paper, he sent a preprint to Alain Léger at the University of Paris. Léger in turn forwarded it to Werner Schmidt at the Institute for Polycyclic Aromatic Hydrocarbon Research in Bayern, Germany. At Léger's suggestion, Schmidt wrote to Wolfgang, suggesting that he attempt to extract the $C_{60}$ from the soot using either sublimation or an organic solvent. In short order, Wolfgang and Kostas succeeded using both methods, and within a couple of days Wolfgang called Don with a succinct message, "Put it in benzene."

**The Main Event**

Wolfgang's call came just a few days before Don's flight to Europe. This was a three-week trip Don felt he could not miss: he was traveling with a colleague and visits to a number of observatories had been set up months in

[1]With apologies to Fermat, the story of Don Huffman's "Moon project" is well worth hearing, but it will not fit in the margins of this account. Hopefully, it will be recorded elsewhere.

advance. $C_{60}$ couldn't wait that long, so he called me in and told me to drop everything else and jump back onto soot making. Before he left we replicated Wolfgang's results, made our first evaporated $C_{60}$ film, and measured the UV-visible absorption spectrum, which displayed the three strong bands responsible for the "humps" in Kamel soot.

While Don was gone I spent most of my time scaling up our soot production and $C_{60}$ extraction process. Don and Wolfgang were talking daily by telephone, and I talked with Don on the phone or exchanged faxes with him, usually once a day. For me personally this was the time of highest excitement. I was one of four people on the planet who knew how to make $C_{60}$, and during those early days everything we learned and every measurement we made was important. I remember walking through the halls of the Physics building one day, on my way to get an IR spectrum of a sample, when I passed Bob Thews, a theoretical particle physicist and Physics Department Head. "Dr. Thews," I blurted out, "We've got $C_{60}$!" He smiled, replied, "Very good," and walked past me. After a few steps he stopped, turned around, and said, "Wait a minute. That's important, isn't it?" For a graduate student it just doesn't get any better.

When Don returned, things really moved into high gear. The publication of the *Chemical Physics Letters* paper on the sixth of July was especially concerning, since it contained the entire recipe for $C_{60}$ except the final "put it in benzene" step. Our great fear was that if *we* had learned how to extract $C_{60}$ from the soot, any "real chemist" could figure it out in no time. Over the next five weeks Don and I measured the UV-visible spectrum, optical density, and density, while Wolfgang and Kostas obtained a mass spectrum and an IR spectrum. Our eureka moment occurred when we submitted a sample of buckyball powder for x-ray analysis and were handed back a beautiful diffraction pattern. We had discovered a new, crystalline form of carbon, which is not something that happens regularly. Not only had crystalline $C_{60}$ never been seen, its existence had not been predicted or even speculated upon, unlike many of the molecular properties of $C_{60}$. In parallel, Wolfgang and Kostas obtained electron diffraction measurements of a single crystal which confirmed the x-ray measurements, and the term *fullerite* was coined to describe the condensed phase of the fullerenes.

By the end of July the experimental work was complete. To the greatest extent possible, all measurements had been made by both laboratories and the results cross-checked. We spent about a week drafting the paper, during which we suffered from the contradictory anxieties that (1) we were about to get scooped and (2) our results were too good to be true. On the seventh of August I faxed the manuscript to *Nature*'s office in Washington D.C. and then called to confirm that the fax had arrived. I was hurried off the phone by a bored-sounding young woman who told me that yes, we have your manuscript, it will be relayed to the appropriate editor for review, and we will contact you once the process is complete.

I give *Nature* full credit for having a quick and efficient review process. The next day, August eighth, I answered the phone in our lab and a gentleman on the other end, who did not sound bored, identified himself as an editor for *Nature* and asked to speak with Don. Not only had our manuscript been accepted, but Harry Kroto, one of the reviewers, had waived anonymity and offered his congratulations. In addition to suggesting a few small corrections regarding the references, Harry asked (via the editor) why we hadn't included NMR data. Don and I were puzzled by Harry's emphasis on this particular measurement. After all, NMR would only show one peak, and matching up whatever the actual peak position happened to be with a theoretical prediction seemed problematic. The IR spectrum, on the other hand, contained four peaks, and even if you couldn't reconcile theory and experiment in terms of the absolute peak positions, comparison of the relative positions offered a stringent test for the presence of $C_{60}$. To us, at least, the IR spectrum was better proof than an NMR spectrum would be. Don told the editor that we had no plans to obtain an NMR spectrum and agreed to proceed with the manuscript as written [8]. I now understand it was the simple presence only one peak that the chemistry community viewed as compelling and the absolute position of the peak was a secondary consideration. As time went on we came to appreciate both the magnitude of what we had passed up and Harry's generosity in offering us the chance to make this measurement first.

The seven weeks between submission of the manuscript and its publication toward the end of September were very busy. We had mailed out a number of preprints and the phone was starting to ring. Two calls stand out in my mind. The first was from Rick Smalley, who offered his congratulations and told us that his group was happily celebrating what they considered to be the validation of five years of hard work. He also had a favor to ask: would Don mind sending him a small sample of $C_{60}$? "You don't know how important it is to just see the stuff." After the call Don went into the lab, evaporated a thin film of $C_{60}$ onto a microscope slide, and sent it to Rick via overnight courier.

The other call was from Rob Whetten at UCLA. A group within the UCLA Chemistry Department had received a preprint and not wasted a second. In record time they figured out how to make Kamel soot and extract the fullerenes, and then they got to work characterizing the material. Shortly before our paper was published, Rob called Don and told him, "Everything in your paper is correct." Although our confidence in our results had been increasing as time passed, this very welcome message erased the last of our fears.

**1991 to Present**

The next chapter in the buckyball story is well known. Don and Wolfgang's method for producing fullerenes was adopted by hundreds of laboratories worldwide and fullerene science entered the mainstream. At Arizona we received much-needed help from colleagues in the Chemistry Department (for one thing, they switched us from benzene to toluene) and from colleagues elsewhere. I graduated in 1991 and stayed on as Don's postdoc, and for the next five years I had the good fortune to participate in a wide-ranging series of fullerene collaborations. In 1996 I left academia and moved with my family to the San Francisco Bay area, where I have worked in the telecommunications industry ever since.

The fullerenes were a gift. As a good friend of mine put it, I am one of the lucky few who "touched the magic" during his scientific career. I worked with some of the top people in the world on problems drawn from physics, chemistry, astronomy, materials science, and engineering, and I'm well aware this doesn't happen for everyone. Out of all of it, the most valuable lesson I learned came from watching Don and Wolfgang spend five years carefully and methodically untangling the Kamel soot mystery. $C_{60}$ was one of the hottest topics in the physical sciences during this period and there was a very real danger of being scooped at any time, yet they only published when they were certain of what they had. It's a hard lesson to learn, but no matter what happens, and no matter what the external forces are, you have to get the right answer.

9 January 2013

# References

1. C F Bohren and D R Huffman, *Absorption and Scattering of Light by Small Particles*, John Wiley and Sons, New York (1983).

2. K L Day and D R Huffman, "Measured extinction efficiency of graphitic smoke in the region 1200–6000 Å," *Nat Phys Sci*, **243**, 50–51 (1973).

3. "Molecule is shaped like soccer ball," *The New York Times*, December 3, 1985.

4. H W Kroto, J R Heath, S C O'Brien, R F Curl, and R E Smalley, "C₆₀: buckminsterfullerene," *Nature*, **318**, 162–163 (1985).

5. R C Haddon, L E Brus, and K Raghavachari, "Electronic structure and bonding in icosahedral C₆₀," *Chem Phys Lett*, **125**(5,6), 459–464 (1986).

6. W Krätschmer, K Fostiropoulos, and D R Huffman, "Search for the UV and IR spectra of C₆₀ in laboratory-produced carbon dust," in *Dusty Objects in the Universe*, ed. A Busoletti and A A Vitone, Kluwer Academic, the Netherlands, 89–93 (1990).

7. W Krätschmer, K Fostiropoulos, and D R Huffman, "The infrared and ultraviolet absorption spectra of laboratory-produced carbon dust: evidence for the presence of the molecule C₆₀," *Chem Phys Lett*, **170**(2,3), 167–170, (1990).

8. W Krätschmer, L D Lamb, K Fostiropoulos, and D R Huffman, "Solid C₆₀: a new form of carbon," *Nature*, **347**, 354–358 (1990).

# A10

# A PhD student's account of the $C_{60}$ story

## Jonathan Hare

*Physics and Astronomy, University of Sussex, Sussex House,*
*Falmer Brighton, BN1 9RH, UK*

j.p.hare@sussex.ac.uk

**Figure 1** A magenta solution of $C_{60}$ (left), the black soot-like $C_{60}$ containing material produced in the carbon arc (middle) and a red solution of $C_{70}$ (right).

I received my PhD in chemical physics in 1993. The period from 1990 to 1995 was a wonderful and very exciting five

*$C_{60}$ Buckminsterfullerene: Some Inside Stories*
Edited by Harry Kroto
Copyright © 2015 Pan Stanford Publishing Pte. Ltd.
ISBN 978-981-4463-71-3 (Hardcover), 978-981-4463-72-0 (eBook)
www.panstanford.com

years when I was lucky to be involved in the "C$_{60}$ affair". I feel very privileged to have been involved, to have been able to play a small but significant role in the C$_{60}$ story. I worked in a dynamic, happy and very creative team. I felt that I was not only valued but that I also contributed uniquely to the "life" of this group. It was an exciting time.

**Flashback**

It's sometime in 1990, about 6 a.m. on a bright summer's morning. I am in the process of trying to wake up, washing myself in a bathroom above the labs. I have just had a few hours' sleep after spending most of the night in the lab making C$_{60}$. I am one of the few people in the world who can make it. At this critical time I need to "crank the handle" of the apparatus to make as much C$_{60}$ as I can. I find that everything I make gets given away to another expert. I don't always need to work through the night but right now we never seem to have enough of the precious material. It seems that everyone wants some to investigate

**Figure 2** Jonathan Hare in the lab with a fullerene arc apparatus, ca. 1991–1992.

its properties. It's an exhilarating time for me. Everything feels extraordinary—the work has got a spirit to it.

Working late in the labs is lonely but exciting. By 10 p.m. everyone has gone home and you are left alone with the sound of the air ventilation, dry box, fans, motors and pumps. The equipment I am using is home-built and even when it's working perfectly I find I am always "tweaking" it, trying to get better results and to understand each bit better. I don't rush. I find it satisfying to make each batch.

While the vacuum equipment is pumping down, getting ready for the next experiment, I go into Harry Kroto's office to read the latest paper he's working on. He often leaves a printed version on his table so he can immediately get started on it first thing the next day. It's a privilege to read these as I get a day-by-day view of alterations, corrections and developments in his thinking—as I get a snapshot glimpse of what's going on in his mind as the scientific research progresses.

**Figure 3** Harry Kroto in his office (1990s).

The relationship between the PhD student and the supervisor is a unique kind of apprenticeship. So sitting at his desk in the middle of the night reading his ideas and thoughts was a very special privilege. It is intensely interesting to see the day's problems discussed and projected on these latest papers. I realise these are special times, unique times. I savour them. Everything feels a part of something bigger going on in the world.

At about 2–3 a.m. I roll out the sleeping bag and try to ignore the background sound of the labs that now seems very loud. I manage a few hours of sleep on Harry's office floor. In the morning Harry gets a call from the lab manager saying I should not work alone in the labs. *God, if they knew I had slept here!* Later, Harry tells me he used to sleep in his supervisor's office when he was doing his PhD.

## The Scientific Background

In 1989 W. Krätschmer, F. Fostiropoulos and D. Huffman published a fascinating paper in the proceedings of an astronomical conference. They claimed to have found C$_{60}$ in a soot-like carbon layer deposited by arcing carbon rods in an inert gas atmosphere (helium or argon). Theoretical analysis predicts that C$_{60}$ should absorb infrared light in only four places (4 IR bands). These soot films showed the presence of four weak bands with intensities that suggested it made up about 1% of the soot. On the basis of these results the authors claimed that C$_{60}$ could therefore be made in this very simple way. To those not involved in the field it may come as a surprise to hear that this paper was almost completely ignored, probably for the following reasons. Estimates of the original discovery cluster beam data suggested that the amount of C$_{60}$ being formed from the graphite was only about 1 part in 10,000 at best, and probably more like 1 part in million. If one could obtain only this sort of production in the relatively sophisticated cluster beam equipment, one would really not expect a simple carbon arc to be efficient at all. Further, a cage molecule, $C_{20}H_{20}$, dodecahedron, had been made in the lab, in small quantities by a 16-step synthesis from suitable precursors. $C_{60}$ (roughly three times the size) was therefore expected to be a difficult molecule to synthesise. Despite these reasonable scientific problems, the Krätschmer group claimed that their technique was not only incredibly simple but also produced relatively high quantities of $C_{60}$.

## Degree to PhD

My first degree was in physics at Surrey University. Before I had finished my degree at Surrey I wrote to Sussex University, enquiring about a PhD in astronomy. I had a very positive interview at Sussex with Robert Smith. A few weeks later I received a letter from him suggesting I contact Prof. Harry Kroto in the School of Chemistry and Molecular Sciences. Gill Watson phoned me up at Surrey and arranged for me to come down to Sussex University to meet Harry.

One bank holiday he picked me up from Lewes station and, going around with Harry, I remember how even the town seemed studious. Harry showed me around the labs and told me all about this fascinating new form of carbon: $C_{60}$. He showed me the recent data he had brought back from the US experiments; they looked strange to me. On reflection it was because they were digital rather than analog signals. In other words, they were collected by a digital computer and printed out, they were not smooth, they were jerky and odd-looking. Not thinking properly, I said to Harry, "Are these real data?" He looked a bit hurt and said, "Of course, they're REAL. What do you mean!?" The research seemed so fascinating, and Harry was so enthusiastic that I really wanted to work with him.

## A New Postgrad: The First Experiment

I started with Harry on 10 October 1989. He was keen to try and reproduce the fascinating results of Krätschmer and Huffman's group. Two years earlier Harry's group had made similar materials by the same method but had not continued the investigation because of lack of financial support for an in situ mass spectrometer he thought would be needed to detect and monitor the minute amounts of C$_{60}$ he expected to form in this simple system. Harry considers this by far the biggest scientific mistake he ever made!

**Figure 4** Amit Sarkar and Jonathan Hare with the old bell-jar apparatus.

The old, worn-out carbon arc vaporiser used in these earlier experiments was resurrected and Simon Balm (also a DPhil student at the time), and I spent a few days playing around getting it working. Harry had originally drilled a gas inlet hole into the base of the vaporiser to allow an inert gas into the system. By 17 October we vaporised carbon and obtained a deposit. Then Simon left to work on the cluster beam apparatus in another part of the building and I was joined by Amit Sarkar, a final-year chemical physics undergraduate student. We worked together in order to try and reproduce the results of the Krätschmer paper, which would also form Amit's final year project.

During the next few weeks we found that the nature of the carbon deposited was critically dependent on the gas pressure under which the rods were vaporised. On 22 October we confirmed the IR results for the first time. However, we could not reproduce these results consistently. During the following week, after continued use, the old equipment could take no more; the insulation on an old Variac failed and burned out, the wiring left a lot to be desired and the pressure gauges and pump connections needed changing.

Harry came into the lab and said, "You just start to get some results and the apparatus gives up, typical!" I spent the next week rebuilding the vacuum system and the electronics. Steve Wood (my British Gas award supervisor) brought me a welding kit power supply to replace the old vaporiser transformer. When the apparatus was finally in working order we were allocated a small disused darkroom in which to set it all up.

By the 11th of December we had conducted more experiments and we were able to reproduce the spectra a second time. It really did look as if the soot contained something interesting. Amit and I continued the experiments but we could still not get reliable spectra from the films; only about one in every ten produced a positive result.

Amit finished his project with two positive IR spectra confirming the results of Krätschmer. Meanwhile I pressed on, trying to make reproducible films so that I could pinpoint the exact conditions which would give rise to the four bands. To do this I varied every parameter I could think of while trying to keep all else constant.

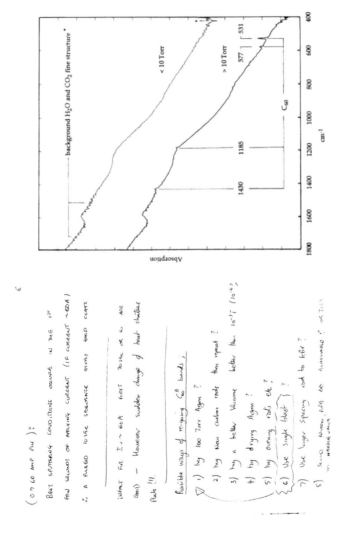

**Figure 5** Extracts from Jonathan Hare's lab notebooks. *Left*: Various experiments to adjust the quality of the $C_{60}$ containing films. *Right*: Typical IR spectra showing the all important four IR absorptions we hoped were $C_{60}$.

While I was conducting these experiments I was also building equipment for the cluster beam experiments. On 28 February 1990, I wrote to Krätschmer telling him of the results we had obtained and received a friendly, positive reply.

**A Few Extracts from my Notebooks**

*07/03/90 – tried moving the [collecting] plates back to ~60mm – seen $C_{60}$ again! I repeat the experiment. It's a better coat and seen it even stronger! I am pretty sure now that if I can get a decent coat (which should be quite possible now I have a new power supply) I will see $C_{60}$. Mind you I've thought that before! So I'm not counting on it completely. However I've eliminated many possibilities and 'niggles', so my mind is quite at rest over the matter.*

*09/03/90 – A good day. Repeated $C_{60}$ expts. – it worked! Even better than before! I found that the coats on the KBr [collecting plates] can't-be-too-thick. Also tried helium instead of argon and much the same results.*

*14/03/90 – really tired by the end of today and the interesting results unfortunately had the opposite effect on me and made me feel a bit fed up – I was expecting a good steady increase – no nature wouldn't allow that! Too simple. Oh well, sort it out tomorrow when I am not so tired.*

After some time I found that the separation of the depositing surface from the arc was the important factor once the pressure of the gas was greater than the value noted by Krätschmer. This allowed me to detect the bands routinely and thus to see, for example, how gas pressure or separation affected the spectra. (With hindsight, it is probable that in the case of films deposited too near the hot arc, most of the fullerene content may sublime away, and therefore no IR bands are detected.)

The results of these measurements were presented at a conference on molecular clouds in Manchester on 25–29 March 1990. A lot of interest was shown and several reprints were requested. One encouraging comment came from the astronomer Patrick Thaddeus who said, *"I think you really have got something there!"*

The next few months were spent analysing some data obtained by the Giotto space probe, which passed through the tail of Halley's Comet (this became my first scientific paper). As time permitted, every few days I "fired up" the vaporiser and made more soot. Every time the correct spectra were recorded, the soot was scraped off the walls and collected in a bottle. In this way I managed to collect a small stock of the precious (C$_{60}$-containing?) material.

By the 29th of May we thought we might have enough to run a technique called solid state NMR—which should have been able to help identify the material. Three samples were submitted, so that any differences in the materials could be recorded. Unfortunately the NMR equipment had a faulty spinner and there seemed to be some confusion over the timing pulses needed to probe such samples. Consequently, no results were obtained.

The Halley Comet paper was written up by 12 July and sent off. I was looking forward to a holiday in Scotland the following week, but before I left a friend of mine, Nick Blagden (also a postgrad), suggested that I submit a sample of the soot for mass spectrometric analysis. I gave a sample to Ala'a Abdul-Sada, a postdoctoral fellow who has great expertise in mass spectrometry. He seemed very keen and said he would run the sample as soon as possible.

The following week was spent walking with my friend Guy Noble through 80 miles of the Scottish mountains along the beautiful West Highlands Way, in gorgeous sunshine. This was a great experience and a time for a well-earned break. When I returned to Sussex, Ali immediately contacted me saying he had some very exciting results.

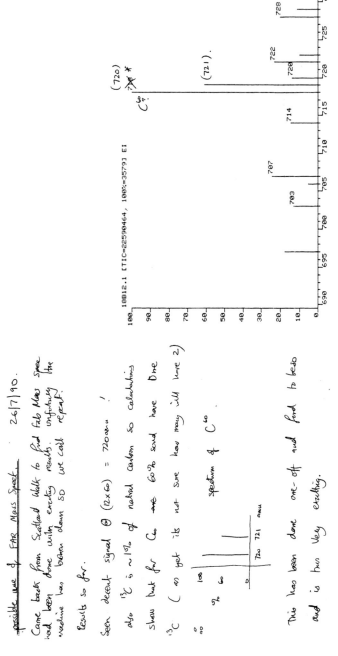

**Figure 6** Extracts from Jonathan's lab notebooks. The first mass spectrum of $C_{60}$ from carbon films generated by the carbon arc apparatus.

3/8/90

1.) Made aprox ½ a (30mℓ) tube of $C_{60}^B$ + Carbon Powder, Actual Volume would be much smaller than this b'cus powder is so uncompact.

2) added about 25mℓ of Benzene and shook mixture

3) allowed to stand for weekend.

6/8/90

Solution looks slightly redish, tried to pipet liquid out from top but mixed up.

9/8/90

Vacume lined sample to about 5th of Volume could go lower (ie more concentrated) but we need about this Volume if we want to use IR liquid cell, so will keep to this.

Continued evaporation down to about 4-5 drops (1mℓ?). FAB showed No $C_{60}$ (720).

While Lumy was Vacuming to get Benzene off I noticed that the liquid froze early and produced beautiful hexagonal crystals interlayered other fumes

were probably produced but I could not tell there shape ie sort of amorphous + toluene fumes was also produced, but probably was above but in cooling was disturbed by movement.

**Figure 7** Extracts from Jonathan's lab notebooks. Adding benzene to the $C_{60}$ containing soot and getting a red solution. *Bottom*: Pure $C_{60}$ in toluene.

However, in his own words, "we have peaks at the correct positions Jon, but the machine has broken down so we can't repeat the measurements" (!). The spectrum showed the peak profile and isotope intensities expected for $C_{60}$ (but the calibration was in error by 2 mass units). The importance of this spectrum cannot be over-emphasised, because it was the first experimental evidence, apart from the IR results, to confirm the presence of $C_{60}$ in the soot.

## The Red Solution

On the 26th of July I read another important Krätschmer et al. paper showing that pure $^{13}C$ (a very expensive "heavy" form of carbon) gave the same four-band IR spectra but with the expected shift in frequency. This showed that the spectra originated from carbon and not from "junk" evaporated into the arc, and also that the molecule was carbon-rich. Krätschmer and his co-workers also referenced our findings in their paper—which was very kind of them, and exciting!

This paper was very important because it reminded me that the world was moving on out there and that someone might actually make $C_{60}$ very soon. This niggling worry didn't help things but it pushed me along.

On the following Friday I thought again about trying to extract the $C_{60}$ from this soot. Over the past months I had made about 10 mg of soot which I knew contained the substance that produced the IR bands and the 720 amu mass spectrum peak. To try and extract $C_{60}$ seemed the next logical step. Benzene seemed the obvious choice. I took about half of the soot I had made and added it to about 20 ml of benzene in a small vial. I put it on the shelf and left it over the weekend. To be honest, I didn't expect much to happen, but it seemed well worth a try. When I came into the lab on Monday morning (6 August) the soot had settled and the benzene had turned red!, albeit rather faint.

On filtering and evaporation, the liquid colour became far more intense, as one would expect. I went round the lab proclaiming (though somewhat half-heartedly!), "C$_{60}$'s in here!" shaking the small vial in front of people's eyes, "Yes, OK Jon...," but it turned out that this was exactly what it was.

A red solution (or any other colour for that matter) from pure carbon was a totally unexpected result. In fact the solubility of C$_{60}$ remains one of its most important properties, allowing chemistry to be carried out straight-forwardly. Later we were to show that one could extract a whole family off fullerenes just by washing the soot in benzene (or toluene).

Harry's response was one of excited caution. We were both worried that the colour may be due to a colloidal suspension of tiny soot particles—we must obtain a mass spectrum from the liquid, but because of yet again malfunctioning equipment at Sussex, the spectrometer gave no signal.

### The Birth of Fullerene Science

On the morning of 10 August, Harry asked me to go down to the chemistry stores and pick up a fax that had just come through from the journal *Nature*. On the way back I nosed through the papers intently. By the time I reached Harry's office I could tell he must have known what was in the paper. In the following silent minutes, which seemed to last forever, I watched as Harry flicked through the pages with such concentration.

*Oh my God, they've done it!*

The *Nature* paper was great! They had made the soots in the way described above and obtained a solid from it which showed, by IR and UV spectroscopy and X-ray diffraction, results consistent with a molecular structure of small balls, *and* it was found to be soluble in benzene to give a red solution (!). My feelings were mixed. On the one hand all this work I had been doing was not in vain; $C_{60}$ really did exist in the soots, which was rather satisfying. On the other hand it looks like we had been so close! My view is that we could have proved the structure of $C_{60}$ at Sussex but I would have done it cautiously in my own time, probably in the remaining two years of my degree. However, this was not to be the case because the fullerene story had entered a new phase.

After what must have been an indigestible lunch, Harry sent a fax back to *Nature* expressing his congratulations to the authors. Soon everyone in the School came to know about the discovery. I meet Chup Yew Mok, a visiting professor. He said, "Jon, everyone in the whole department is talking about your work!"

The *Nature* paper described a classic example of a new discovery in science and because of this it deserves all the fame it has received, and rightly so. However, the next crucial step in the story was carried out here at Sussex with Roger Taylor joining the team. The next day Roger came up to talk to Harry. He had some ideas on how to extract and possibly separate the material more effectively. He asked for all the material I had made. I was very reluctant to let this precious stuff go, so initially I gave him half my total stocks. Meanwhile I pressed on making as much material as possible. Roger used the Soxhlet method to extract much more material from the soot. We got an 8% yield of $C_{60}$!

However, a really exciting discovery came when Roger, with advice from a Sussex colleague, Jim Hanson, chromatographed this extracted material. Chromatography is a very powerful technique used to separate mixtures. He used the solution from my soot samples and obtained at least three coloured bands of material separated on the

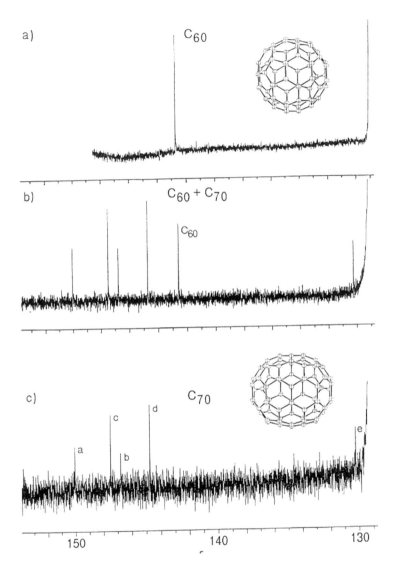

**Figure 8** NMR resonances for $C_{60}$, $C_{60}$ and $C_{70}$ mixture and pure $C_{70}$. $C_{60}$ gives a one-line resonance, while $C_{70}$ produces five resonance lines in the ratio 1:2:1:2:1.

chromatography column. One was the characteristic red colour but the first fraction was magenta. What was this first fraction?

One result from the initial 1985 discovery was that whenever $C_{60}$ formed, then so should $C_{70}$ (like a Rugby ball!). Could this first band be $C_{70}$? The *Nature* paper fax focused only on $C_{60}$. The one measurement that would prove the structure without doubt was a technique called $^{13}C$ NMR. There were no NMR measurements in the *Nature* manuscript or mass spectrometric data. If we could record this, all would not be lost! Gerry Lawless and Tony Avent of the Sussex NMR lab ran these fractions for solution NMR. Because all 60 carbon atoms in $C_{60}$ are equivalent, the NMR spectrum should consist of only one line, and if $C_{70}$ had the proposed structure, then it should give a five-line spectrum with predicted relative intensities 1:2:1:2:1 in some order. The results were exactly as expected. We were the first group to measure the NMR of pure $C_{60}$ and $C_{70}$ and prove their structures unambiguously by this powerful method. All was not lost!

Around this time we had lots of visitors to the laboratory. Elaine Seddon came over to run a technique called PES spectrum on gas phase $C_{60}$. The results of this measurement compared to astronomical spectra might help to see if $C_{60}$ was "out there" in space. I worked with her through the night till 8 a.m. for a few days running. Roger at this time was like a five-year-old at Christmas constantly grinning all the time. The football team, as Harry called us, was in action!

To be as versatile as possible we used to do all sorts of experiments to see if anything interesting would turn up. On one occasion I used $SF_6$ instead of helium. We hoped that it might make a fluorinated analogue of $C_{60}$, perhaps $C_{60}F_{60}$ or even a ball with a fluorine atom inside it—called an endohedral fullerene. I set up the apparatus as usual, introduced the new gas and struck the arc. When I opened it up, instead of seeing lots of black sooty powder I found that there wasn't anything to see at all. On closer inspection I found that all the surfaces inside were covered in a brown plastic-like layer. I had in fact made the fluorinated carbon polymer—Teflon—the same stuff that non-stick pans are covered in!

We were still having problems with the mass spectrometer at Sussex, so I took the two (now very, very precious) samples up to VG Analytical in Manchester to run on their superb machines. Harry, Roger and David Walton, who had originally initiated the carbon chain research that had led to the present work with Harry several years before, waited by the phone at Sussex for me to ring them with any news. After much effort we did eventually record the mass spectra we needed.

We discovered that C$_{60}$ is magenta in solution and that the red colour seen in the first extract was in fact due to the more strongly coloured (but less abundant) C$_{70}$. These results were published very soon in our paper sent to the journal *Chemical Communications*. Roger continued the chromatography and I converted an old pyrolysis chamber into a more efficient fullerene soot generator.

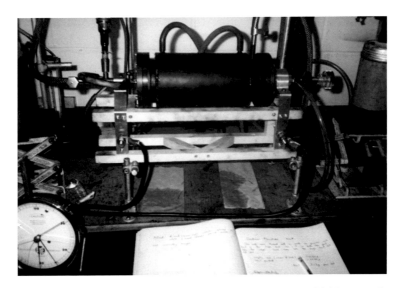

**Figure 9** An early carbon arc evaporator, which gave a 10-fold increase in production compared to the bell-jar apparatus.

## Diary Entries from the Time

*05/09/90 – it's 10 pm and I am still at uni. Been making some carbon batches on the new machine really good results. Must have made 5g of soot that's about 10x what I've made so far! Also the*

**Figure 10** *Top*: The Kroto contingent at the 1996 Nobel Prize ceremony in Stockholm, Sweden. *Bottom*: Harry Kroto receiving the Nobel Prize.

*results of this soot look like they are the best so far, excellent. Got a copy of the paper to be sent to Chem. Comm., looks good – should be in by Xmas (yes, I know it sounds ages but apparently that's fast!). Nobel prize for Harry!? God knows.*

*09/09/90 – Brioni Island Conference. Harry comes back to my room and we chat about C$_{60}$ and C$_{70}$ etc. gave him the spectra and overheads he was very chuffed and very pleased and said very nice things about my work etc. He also realised he had been hard on me the last few weeks – nice guy. It was magic moment and we both realised that the C$_{60}$ affair couldn't have happened the way it has without both of us – so I felt very important and happy! Sat on the pier alone and felt really happy inside, thought of Gaia etc. what a wonderful Earth. Brioni is so beautiful.*

## Harry Kroto's Nobel Prize, 9 October 1996

On 9 October 1996 Harry, Bernd Eggen, Gill Watson and I went for lunch at the Sussex Gardener Arts Centre. Just before we went to lunch, Bernd checked the Web for news of the Nobel Prize. However, they did not give out the news till later, so we left it and went to eat. At lunch we chatted about who might get THE prize and Harry suggested Ahmed Zewail and Dick Zare. He wanted to superimpose their faces on a picture of the Texas rockers ZZ-Tops. I remember listening to the conversation with some interest but I was actually happy to eat my cake and drink the coffee. I wasn't thinking about the possibility that Harry might win, although it had been mentioned in previous years.

After lunch we came back to the lab and Bernd looked up the latest info on the Web. The three of us stood there waiting to see who had won. As the page came into view, Bernd gave a mighty yell and ran out of the room. Harry and I looked at each other a bit puzzled trying to work out why Bernd was so excited. The words on the screen seemed to be hidden from us, then we both saw it: HARRY KROTO! Bernd came rushing back into the lab with a bottle of champagne and I got some glasses. Harry and I were both a bit dazed— was it really true?

**What we read on the website**

"The Royal Swedish Academy of Science has decided to award the 1996 Nobel Prize in Chemistry jointly to

Professor **Robert F. Curl, Jr.**, Rice University, Houston, USA,

Professor **Sir Harold W. Kroto**, University of Sussex, Brighton, U.K., and

Professor **Richard E. Smalley**, Rice University, Houston, USA,

*for their discovery of fullerenes.*"

**The Answerphone Recording**

The next thing I remember was, *Must check the answerphone.* On the machine was a collection of messages, one of which was from the Royal Swedish Academy, a kindly but sober voice informing Sir Harry that he had been chosen as a joint winner of the 1996 Nobel Prize for Chemistry. Soon after listening to the message on the answerphone, I rushed around playing it to people. I ran into David Walton's office without knocking and said, "Listen to this!" I can remember how Dave's eyes lit up. Almost immediately the phone calls and faxes started coming in. It felt like a number of birthdays and Christmas days rolled into one!

# A11

# The C$_{60}$ buckminsterfullerene formation process: new revelations after 25 years

## Paul Dunk

Department of Chemistry & Biochemistry, 95 Chieftan Way Rm. 118 DLC,
National High Magnetic Field Laboratory, Florida State University,
Tallahassee, FL 32310, USA
pdunk@chem.fsu.edu, dunk@magnet.fsu.edu

### Starting Out

In 2005, I met Harry Kroto for the first time in his office at Florida State University's chemistry department in Tallahassee, Florida. I was finishing up my undergraduate degree in biochemistry and had been performing under-graduate research in synthetic organic chemistry for almost two years up to that point. During that time, I had become relatively well trained in synthetic techniques, developed a chemist's touch in the lab, and learned to enjoy the challenge of research. I had planned to

C$_{60}$ Buckminsterfullerene: Some Inside Stories
Edited by Harry Kroto
Copyright © 2015 Pan Stanford Publishing Pte. Ltd.
ISBN 978-981-4463-71-3 (Hardcover), 978-981-4463-72-0 (eBook)
www.panstanford.com

pursue graduate studies in the field of synthetic organic chemistry. Harry set up that meeting with me after I contacted him to see if we could chat about what research projects he was into at the time. For some rather traditional organic chemists, it was perhaps viewed as a waste of time to work on projects outside of the area, particularly when graduate school was approaching. But had I realized during my undergraduate years that personal growth and great things often result from moving outside of one's familiar surroundings and comfort zone, even though it's almost never easy.

It was a bit surprising that Harry would take the time out of his hectic, globe-trotting schedule to meet with an undergraduate about doing research. And I was pretty sure he was not going to be impressed with my knowledge of nanoscience; I knew about C$_{60}$ primarily because it was on the cover of my undergraduate chemistry books. But soon into the meeting, it was very clear that he didn't look down on anyone. Instead, he appeared to enjoy getting students into research. He suggested several topics and possible areas to work on, and wanted to make sure that the project was something that captured my interest. I looked into the suggested areas, read some relevant publications and book chapters, and after some contemplation, settled on carbon nanotubes. In particular, I thought it would be interesting to see if we could develop a method to better control carbon nanotube formation. And very soon after, we started the project.

Some intriguing results were obtained, although they were rather difficult to interpret. In the course of that work, Harry encouraged me to apply for an internship program for young scientists at a fantastic research and microscopy center: the National Institute for Materials Science located in Tsukuba, Japan. I subsequently received the internship, which was wonderful because I could now develop the experiments at a global carbon nanotube research hub and analyze the results with some of the best transmission electron microscopes in the world. The experience of working at a national laboratory in an international environment turned out to be an extremely important and useful experience for our later work on deciphering the long-standing puzzle of $C_{60}$ formation.

## Research in a Foreign Country

After a 13-hour flight, I arrived in Japan, then started my journey to the city of Tsukuba by train, then bus, and finally taxi. The scenery of Japan was amazing. Upon reaching the apartment housing for researchers, I was quite surprised to see that the Japanese taxi driver utterly refused a tip that I attempted to give him. In order to make the trip, I had to finish my final undergraduate courses before the semester was completed. Therefore, there was very little time to learn the Japanese language or look deeply into the culture before I left. It became very clear, however, that understanding how best to interact with, know, and appreciate other cultures was going to be very important. With extreme jet lag, I woke up the next morning and attempted to make my way to the laboratory. It turned out there were several NIMS buildings in Tsukuba, and I walked to the wrong one. Then, I got completely lost on the way to the correct building. I finally reached the institution, with bleeding ankles that stained my pants from walking for what seemed like miles in shoes not meant for it, and met with the director and the institution members that I would be working with over two hours late. The rest of the day was spent trying to stay awake and take in all the information that was available to me.

On my second day in Japan, we started research. Indeed, one reason why was I able to start so quickly was because of the assistance and mentorship of a Russian co-worker. Figuring out how to design and perform new experiments,

**Figure 1** Clockwise from top *left*: (1) field emission transmission electron microscope at NIMS in Tsukuba, Japan, in 2006. (2) High-resolution TEM image of a carbon nanotube tip produced during research at NIMS. (3) The Ginkakuji temple in Kyoto during the summer. (4) Picture taken of the temple-filled, beautiful city of Kyoto during a visit.

as well as learning everything about the other instruments and microscopes at the facility (**Fig. 1**), was rather dizzying. But with that stress, there was also much excitement about starting the research and meeting the many other young, international researchers at the institution.

I soon learned that a useful part of the program was a morning coffee break in which those who could pause research would come together to discuss projects and get to know one another better. Nearly all of the researchers had recently received their PhDs, as well as one or two that were in late stages of PhD work. Because I hadn't even started graduate school yet, it was rather overwhelming to discuss research with such an experienced group of peers. However, they were all very kind and I soon came to know many of the fellow researchers, who were from Italy, Switzerland, Germany, United Kingdom, Iran, India, Spain, Belarus, South Korea, Thailand and many more countries. It was captivating to see that they, too, approached their research with much enthusiasm and energy. But what impressed me the most was how they designed and performed their own research projects, and then collaborated with one another in a multidisciplinary way to achieve goals that would not be possible individually. Everyone seemed to have knowledge in many areas which allowed each other to communicate across a broad spectrum of science and move forward on new endeavors. Furthermore, they enjoyed discussing science and research with those outside their particular specialty, which was very beneficial.

My research project didn't work out well as I would have liked, but some interesting results were obtained in the end (**Fig. 1**). Trying to gain experimental insight into carbon nanostructure formation is an extremely difficult task, and carbon nanotube formation remains not well understood to this day. For me, however, the experience was a great first step into independent research. I fully realized the level of dedication and commitment required to perform successful research. I also learned not to be hesitant with regard to working on a new project or in a new area that is outside what one already knows; instead, one must rise to any challenge and give it one's best effort. Harry also suggested that, if I was able to leave the lab for a bit, it would be good to see a special place in Japan and specifically mentioned Kyoto, since I had the opportunity. On his advice, I made a trip, which I would likely not have done otherwise, to the beautiful city of Kyoto toward the end of my stay and marveled at the stunning temples and shrines there (**Fig. 1**). That little break from research permitted me to fully appreciate and highly value the country and its culture. While much time was spent in lab and wonderful experiences were made there, I also cherish the time spent with my international co-workers and new friends at various Japanese restaurants, karaoke establishments, bars, and Saturday nights in Tokyo, which is only 45 minutes away from Tsukuba by rail. Science is the same regardless of which country one comes from; it brought us all together to achieve something that we couldn't achieve individually. Personally, it was extremely important to know that so many free-thinking, kind, and talented human beings exist in every country on Earth.

### Enter the Fullerenes

Upon returning to the USA in late 2006, I initiated my graduate studies in chemistry in Harry Kroto's group at Florida State University. Since C$_{60}$'s discovery in 1985, Harry always wanted to carry out follow-up work into discovering the structure of the smallest fullerene detectable; he long ago predicted that a $T_d$ isomer of a tiny C$_{28}$ cage would be a feasible candidate.[1,2] However, C$_{32}$ was the smallest species to have been reliably identified. In general, fullerenes smaller than C$_{60}$ possess much angle strain due to the pyrimidalization of carbon atoms in the structure. Thus, small fullerenes are quite reactive and difficult to study. His suggestion was to study the fullerenes in the gas phase and then "pick out" any C$_{28}$ clusters that formed and react them with hydrogen. If the C$_{28}$ species is the $T_d$-C$_{28}$ fullerene isomer, a distinct addition pattern should be observed: H addition to the four most strained C atoms. Those four strained C atoms are each at the vertex of a tetrahedron and a sort of tetravalent superatom could form that may be quite stable or at least exhibit metastability. It seemed like an intriguing idea that could lead to other things. But, first, I needed an instrument that could make and analyze carbon clusters in the gas phase. Fortunately, Harry had wanted to start collaboration with another member of FSU's chemistry department, Alan Marshall, who co-invented a powerful analytic technique called Fourier transform ion cyclotron resonance (FT-ICR) mass spectrometry[3] and had since the original breakthrough assiduously honed it into one of the most powerful of analytical techniques. The ICR group is housed at the world-class National High Magnetic Field laboratory, located close to the FSU campus in Florida. The FT-ICR mass spectrometric technique provides ultrahigh resolution by use of strong magnetic fields, and it provides unprecedented mass resolution facilitating the analysis capability necessary for the carbon cluster experiments we envisioned.

Alan Marshall enthusiastically agreed to the collaboration, and generously permitted us to use one of his state-of-the-art FT-ICR mass spectrometers and work with his group. On my first official visit, after navigating through a maze of pathways, gigantic magnets and elaborate instruments, I entered the ICR high bay and saw the system that I would be using. It was the first in a line of three instruments in a massive room that contained two 9.4 tesla and 14.5 tesla magnet systems. The superconducting magnet that I would be using was the size of a small car. In addition, an enormous 21 tesla system will soon be up and running as well.

After I finished my first- and second-year requirements and several interim research projects, I began research at the Magnet Lab in 2008. I felt very lucky to be working in such a unique and world-renowned facility. The ICR group contained a rather large, organized team of extremely talented people. I began working with a newly hired postdoc, Nathan Kaiser, who had just received his PhD in chemistry and had worked extensively on ICR instrumentation. He passed on a mass of knowledge, which cannot be found in any book or publication, and helped me greatly to ascend the steep instrumentation learning curve. We needed to build a carbon cluster source and attach it to the existing mass spectrometer to enable the carbon cluster studies. From the lab's connections we received a preconfigured cluster source: AT&T Bell Laboratories was getting rid of a cluster source and kindly dumped it down to the Magnet Lab. It seemed like a great opportunity to jump start the work and an offer too good to refuse, so we gave it try. We were able to get the instrument working, but not exactly how we wanted. In fact, regardless of the modifications that were made, its fundamental design only seemed to produce large carbon clusters. Therefore, there was no chance to study the small carbon clusters we were interested in. Instrumentation can be a very tough business.

Harry and Alan then asked Mike Duncan of the University of Georgia to visit the lab to give us his opinion of the instrument. Mike is an expert on the pulsed supersonic cluster source technique and, in fact, was a key researcher in the Smalley group at Rice when the instrument that was destined to discover C$_{60}$ in 1985 was developed.[1,4] Mike kindly drove from his lab in Athens, Georgia to the Magnet Lab in Tallahassee, Florida, to have a look at the instrument. It turned out that the cluster source that was given to us possessed a configuration that completely constrained it from being used the way we envisioned. Another lesson I learned: You must become an expert in whatever you do.

Harry would say to me at the beginning of my graduate career, "You are the world's expert in whatever you are doing." That meant, know the background information of your project better than anyone else; know every detail of what you are working on and how your system functionally operates better than anyone else, and so on.

While I had read and deciphered the caveats of over 25 years of fullerene literature, I assumed that the carbon cluster source would work because it was previously built at a large, well-established scientific research company. However, I did not read and critically think enough about the cluster source instrument to realize its potential flaws. Mike Duncan, after pointing out the instrument problems, graciously agreed to have a phone discussion with me after he sent some documents regarding carbon cluster design.[4] That information and subsequent discussions with Mike were absolutely critical in building an effective instrument. If we had not had this expert information, it might have easily added years to the project because we would have had to develop a carbon cluster source from scratch. Thus, discussion and interaction with others is critical for success in science.

**Building the Instrument**

With the designs set and everything planned, we needed all the components to be machined, which, of course, required time. Until the instrument components and parts could be machined or manufactured, I used other techniques to investigate small fullerenes. I spent time in London in the laboratory of John Dennis, a former student of Harry and now professor at Queen Mary, University of London, UK. There I gained experience using an arc-discharge instrument for fullerene synthesis. Subsequently, I returned to Japan again and traveled to the laboratory of Hisanori Shinohara, a long-time friend and colleague of Harry, at Nagoya University, where I used the laser furnace as well as the arc-discharge technique.[5] The experience of working in his top-notch nanoscience laboratory was incredible. We achieved some very fascinating results, but there was still no sign of $C_{28}$. Unfortunately, it appeared that the smallest fullerenes

**Figure 2** The in-house built cluster source and 9.4T FT-ICR MS instrument used for fullerene formation studies, located at the National High Magnetic Field Laboratory in Tallahassee, Florida (USA).

were still too reactive to study by those methods. Nori and his amazing lab members became dear collaborators from that point on. I credit my graduate advisor, Harry Kroto, as well as Hisanori Shinohara, both brilliant scientists and human beings, for a significant part of my development. I cherish the many discussions we had as well as observing how world leading nanoscience laboratories operate.

In 2009, back at the Magnet Lab in Florida, we finally had everything machined and ready. The new pulsed carbon cluster source was assembled (**Fig. 2**), and much effort was required to get the instrument in working order. However, regardless of which parameters we changed, there was no signal. Carbon clusters were not being detected or there was something wrong with the cluster source or analysis instrumentation. At this point, I had seemingly read every paper on cluster sources and fullerenes, and relevant FT-ICR papers. After critically thinking about the many components of the experimental set up, I could not find a flaw in the design of the cluster source, and neither could Nate. Perhaps, there was a mechanical failure in some aspect of the instrumentation? That eventually led us to literally tearing apart the entire instrument over the next three months. Everything we tested appeared to be working correctly; there was no apparent flaw in the overall instrument design. Still, there was absolutely no signal. It eventually got to the point where I would wake up in the morning and could safely predict that after working all day on the instrument, I knew I would be leaving at night a non-functioning instrument. That dreadful feeling lasted for a couple months straight, but we never gave up. Finally, one day the ICR group's software designer, Greg Blakney, had a look at our data acquisition program. After a few minutes of sorting through various files, I heard a loud keyboard tap and Greg said, "That should do it." It was a software issue all along! While I became competent in all aspects of the instrumentation, I failed to closely look at programming issues. That mistake caused months of delay in achieving a working instrument, and is a mistake I would never overlook again. Nothing should be overlooked.

To be sure, the experience of physically examining and testing every component of the instrument turned out to be a gift. It permitted deep knowledge, a strong familiarity with the instrument, and critical experience in instrumentation. I understood and had a working knowledge of everything about it now. What looked like 100's of random wires sticking out of the instrument when I first started, now all had a meaning attached to them and I knew exactly how they functioned. I could envision every component on the inside of the steel chambers, along every inch of the instrument, how they worked together, and its limits. Achieving that level is what Harry meant when he said you must be the world's expert in whatever you do. It felt like we were on the brink of hopelessness that the instrument would never work, but experiencing that pure determination, patience, and hard work could overcome an apparently daunting and a seemingly unsolvable problem was important. Success in research does not come without resilience.

**The Smallest Stable Fullerene (so far!)**

Finally with a working instrument, I was able to delve into the experiments I had been waiting to do. The fullerene $C_{28}$, however, was nowhere to be found. That hypothetical distinctive reaction pattern was not observed; in fact, the smallest fullerene found in any abundance as $C_{32}$. Many different stabilization gases were employed, such as hydrogen, chlorine, and nitric oxide, which I attempted to add during formation or after; but still nothing. The $C_{28}$ cage was just not there. It was my nightmare scenario: taking so much time to build an instrument to study something that may never have existed in the first place.

Placing an element on the inside of the cage, however, was an avenue that facilitated $C_{28}$ formation and detection. A very cool thing about fullerenes is that metals are able to fit into the nanoscale-sized space inside the cage. Perhaps, an encapsulated tetravalent metal could stabilize it rather than addition of 4 exohedral substituents? Metallofullerenes have long been interesting species for which desirable applications as contrast agents in MRI of key elements in renewable energy devices have been suggested, so that experiment seemed quite interesting to attempt. Several years after Harry had predicted that $C_{28}$ should exhibit tetravalent behaviour—indeed it can be considered to be a giant cluster atom analogue of the carbon atom itself—the Rice group found circumstantial evidence that indicated that U@$C_{28}$ was a favored species, thus providing compelling support for Harry's original conjecture that $C_{28}$ should be tetravalent.[6]

Our cluster source was designed to use commercially available, quarter inch diameter carbon rods as the vaporization target. To form endohedral fullerene $C_{28}$, or M@$C_{28}$ (where M is a metal atom inside of the cage), a carbon rod doped with the metal of interest was required. While trying to figure out how to perform the experiment as quickly as possible, I recalled that I saw some manufactured metal-doped rods in the lab of Hisanori Shinohara during my stay in Japan. I asked him for some and he kindly sent them to us. After machining them to the appropriate size for use in the instrument, a flurry of experiments was performed with them. As I tested the various rods for M@$C_{28}$ formation, I got a very exciting result; under certain conditions, vaporization of a Ti-doped carbon rod gave a

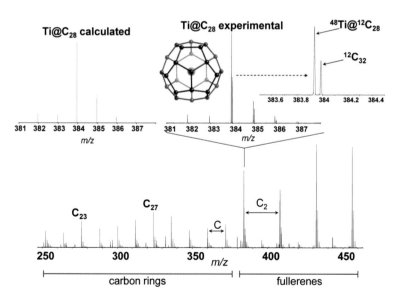

**Figure 3** The tiny $T_d$-$C_{28}$ fullerene is grown a titanium atom. Note that high resolution is needed to detect Ti@$C_{28}$, the smallest fullerene, because $4^{12}C$ is the same nominal mass as $^{48}$Ti.

cluster corresponding to 28 carbon atoms and a titanium atom (**Fig. 3**), which formed as an abundant species.[7] Because these conditions only generated fullerenes, it seemed highly likely that the cluster was indeed the elusive C$_{28}$ cage with a single Ti atom trapped inside. I remember just sitting at the instrument with a giddy smile across my face that I couldn't control, watching the data appear. Caution was still needed because it could still be a different molecular structure, such as carbon ring or linear chain and not a cage. Many additional experiments were performed, including fragmentation and reactivity studies, to elucidate the special cluster's structure. I was able demonstrate that is was a fullerene and so far the smallest one identifiable.

I had exhausted the several special metal-doped rods I had received from Hisanori, but really needed to investigate M@C$_{28}$ formation with many other metals to understand how Ti stabilized the C$_{28}$ cage. First, I tried to find a company that could make metal rods that could fit into my existing instrument. Not a single supplier had such rods. Perhaps I could find some company in the world to make them, but the cost would surely be huge and, importantly, would take a good deal of time even if a manufacturer agreed to make them. That led me to develop a rather convenient method to make my own metal-doped rods and modification of the instrument to permit use of the rods. Now, I could make a carbon rod with any metal I wanted and whenever I wanted. Something so seemingly simple turned out to be critical in moving the project forward.

To further understand how tetravalent metals form the smallest (meta)-stable fullerene, M@C$_{28}$, some theoretical studies would greatly help. Since we did not have anyone at the Magnet Lab or FSU's chemistry department who could quickly tackle such calculations on fullerenes, I needed to find someone who could. I had come across and admired the inventive theoretical papers on metallofullerene written by Toni Rodriguez and Josep Poblet at the Rovira i Virgili University in Tarragona, Spain. In fact they had recently produced several pioneering studies on these systems. After Harry and I asked them if they could help, they agreed, along with their PhD student Marc Mullet-Gas to investigate $C_{28}$ theoretically. Their study indicated that four electrons should indeed be readily donated to the $C_{28}$ cage to form a salt, $Ti^{4+}@C_{28}^{4-}$, which would indeed be expected to be a favored species. Simple chemical structure arguments indicate that four electrons are expected to be required to form a stable, electronically closed shell $C_{28}$ cage. Thus, tetravalent metals would be expected to stabilize $C_{28}$. Experiments were also carried out which gave an intriguing insight into metallofullerene formation. We were able to show experimentally that the encapsulating metal nucleates or in fact appears to catalyze $M@C_{28}$ formation in the gas phase.[7]

## Closed Network Growth of Fullerenes

It was astonishing to me that ever since the discovery of the third form of carbon in 1985 and the subsequent numerous fullerene research studies, the Nobel Prize, and image of $C_{60}$ becoming an iconic image of "Chemistry" and "Nanoscience" known across the world, that some important details of the genesis of $C_{60}$ remained unknown. Thus, understanding how $C_{60}$ forms seemed to me to still be a major challenge. There have been extensive efforts over the years to understand how the perfectly icosahedral $C_{60}$ forms spontaneously in such high abundance. How could this mystery of fullerene formation still persist after nearly 25 years? Because fullerenes self-assemble in carbon plasma, the highly energetic formation conditions and the extremely fast carbon nucleation chemistry which takes place have made conclusive experimental investigation problematical. Regardless of this, numerous fullerene formation mechanisms have been proposed theoretically but all had been based on quite limited experimental information.

Perusal of the literature on fullerene formation leaves one with a sinking feeling that it may remain a mystery for ever. Think of a way to form a fullerene, and it has likely been proposed. The lack of experimental evidence had essentially permitted "anything goes" conjectures: bottom-up to top-down formation and anything in between have been considered. There were few constraints on the conjectured processes that could result in C$_{60}$ formation because too few experiments had been carried to probe adequately fullerene formation in carbon vapour. It was clear that new experimental results were necessary if any useful new insight into the fullerene creation process were to be gained and thus I decided to focus on developing an experiment which might yield an insight into the puzzle of C$_{60}$ formation. I was not that hopeful that much if anything would really come out of this project as there had been so little progress in this area over the last 25 years. I really was not sure that I could figure out a way of obtaining some useful new information on a process that several other groups had already explored.

Fully realizing the experimental challenges in probing fullerene formation, I concentrated mainly on the novel metallofullerene project. The new data we had obtained on M@C$_{28}$, however, really sparked my interest in metallofullerenes and I thought that perhaps some additional new results would be forthcoming in this area. I wanted to know definitively *which metals* would prefer to encapsulate spontaneously inside *which fullerene cages*. Such information could lead to a more detailed understanding of their formation and would be useful for possible isolation of metallofullerenes. The ultrahigh resolution of FT-ICR mass spectrometer would permit the study of any element and overcome the insufficient mass spectrometric resolution that had dogged for previous studies. Due to the many elements of the periodic table and the immense amount of work that would be required, I started to come into lab six, sometimes seven, days per week and work often late into the night and the early hours of the morning.

Many interesting results were coming through all the time. With everyone usually gone, the nights in lab were filled only with the buzz and vibrations of vacuum pumps. Pumping down the instrument after changing target rods gave a little break time in between the hours of long experiments. There was a pond in the back of the Magnet Lab that I would sometimes walk to during such breaks; it was very dark at night as there were no lights in the area but I enjoyed listening to the wildlife. It was a nice substitute for the incessant drone of pumps in the lab. I would often gaze into the sky and appreciate the stars. I would be sure to remind myself that this same starry sky also canvassed over the world in areas where people never had a chance at pursuing things they may like to do or the luxury of worrying if an experiment is going to work or not, as a result of circumstance or of the many other things out of one's control in life. And here I was, at a national laboratory, performing experiments with state-of-the-art instruments. Staying late into the night working on experiments can be a bit lonely, it can be tiring and difficult, it can feel daunting or it can be very exciting, but it is clear to me that, above all, it is a supreme privilege.

After progressing to a little over halfway through the elements of the entire periodic table, and considering the many results on empty cage fullerenes I had obtained under every set of imaginable conditions and experimental parameters, it appeared that fullerene growth could potentially be explained by carbon incorporation into growing cages. To test this idea experimentally, it was clear to me that a fullerene must be exposed to carbon vapour. As I was making great progress toward my goal of testing most of the elements in the periodic table for the metallo-fullerene studies project, the possibility of examining the fullerene formation mechanism further stayed in the back of my mind and I couldn't let it go. Without inventing a new instrument, a potentially very time-consuming and expensive activity that might have only a small chance of being successful, it seemed like one way might be would be to coat the surface of a graphite rod with C$_{60}$ or some other available fullerenes and watch what happened on exposure to carbon vapour in the vaporization source. Carbon vapour would result, as usual, by laser ablation of the graphite rod. The surface-coated fullerene, by contrast, may simply desorb from the rod into that carbon vapour produced from graphite. However, I thought it was unlikely to work well as surely it was so simple an idea it must have been tried before. Furthermore, the pre-existing fullerene would surely be almost completely destroyed by the laser pulse which ablated the carbon. However, because I was now highly familiar with the fullerene literature and had developed a very good understanding of what was going on in our instrument from all our previous results, I decided there was reason for optimism.

I always remembered Harry saying on numerous occasions, "Even if you are sure you know what the result of an experiment will be, try it anyway. Because what actually happens may be unexpected and surprise you." I waited so long to attempt the experiment because I was almost sure I knew what would happen: C$_{60}$ coated on the surface of the graphite rod on laser ablation surely would fragment C$_{60}$ completely or C$_{60}$ would coalesce to form a wealth of larger clusters. I based these conclusions on the fact that if C$_{60}$ becomes sufficiently thermally excited, it fragments by

$C_2$ loss all the way down to the $C_{32}$ cage size. This behaviour, however, was observed by laser irradiation or collisional activation. Further, the timescale to sufficiently thermally excite $C_{60}$ is at least milliseconds and could be possibly as long as seconds in those experiments. By contrast, the Nd:YAG laser I used to ablate a rod had a pulse width of 3–5 nanoseconds. Thus, it is possible that $C_{60}$ would experience a relatively brief period of extreme temperature and could remain stable. The laser ablation of solid $C_{60}$ is known to result in cage coalescence, creating many larger carbon clusters. But for such coalescence to occur, the laser fluence and the $C_{60}$ density must be very high. Surface-coating a rod with a minimal amount of $C_{60}$ would surely yield a relatively low $C_{60}$ vapour density, and the fluence of the laser in our experiments is only a fraction of the laser's maximum output. Hence, our experimental conditions were actually quite different from those in previous studies and it thus seemed to me that $C_{60}$ might not coalesce or fragment in our system and survive to undergo reaction with carbon vapour.

I really did not want to impede the momentum I had developed in other work. But even if the growth experiment didn't work, it would only result in the loss of a day's work. Furthermore, it would not be an expensive experiment to try. Harry's salient advice on trying experiments and these points led me to attempt it. First, I surface-coated a quartz rod with a low concentration of pure $C_{60}$ and then ablated the rod under the conditions I would normally ablate a graphite rod to form fullerenes. The resulting spectrum showed only $C_{60}$ (**Fig. 4**); neither coalescence nor any significant fragmentation occurred. It was a simple spectrum, but it was potentially extremely important! It demonstrated that $C_{60}$ was efficiently desorbed from the rod by the laser pulse with less than 2–3% fragmentation.

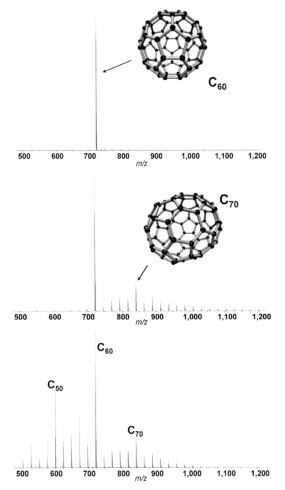

**Figure 4** (Top) Laser-ablated C$_{60}$ in absence of carbon vapour. (Middle) Formation of larger fullerenes after exposure of C$_{60}$ to carbon vapour. (Bottom) Typical distribution of fullerenes formed from carbon vapour, without pre-existing C$_{60}$. After exposure of C$_{60}$ to carbon vapour, larger fullerenes are formed which exactly mimic the >C$_{60}$ distribution when vaporizing pure graphite (compare middle and right spectra). That strongly suggests an identical growth mechanism under both circumstances. Fullerenes smaller than C$_{60}$, however, are not observed after exposure of C$_{60}$ to carbon vapour, indicating a bottom-up growth mechanism. Note that, as shown in right spectrum, small fullerenes are abundant when vaporizing pure graphite without pre-existing C$_{60}$.

This very simple experiment indicated that surface-coating a graphite rod might be an excellent way to "expose" $C_{60}$ to the reactive constituents in carbon vapour. I immediately postponed all other experiments I had planned on in order to probe the consequences of this result. The event reminded me of Harry and his co-workers' discovery of $C_{60}$. Carbon vapour had been previously analyzed by another group, but $C_{60}$ was overlooked.[8] As the Rice-Sussex group looked at carbon clusters formed from their instrument back in 1985, they quickly analyzed critically every aspect of the data they observed. That odd slightly more abundant $C_{60}$ cluster which had been observed by previous workers was probed assiduously, and the rest is history. Now, here another $C_{60}$ peak in a mass spectrum, a simple observation which in my eyes had the potential to lead to some crucial new information and it seemed to me it needed to be investigated immediately.

In the first experiment of the series I surface-coated a graphite rod with the about the same concentration of $C_{60}$ as in the trial experiment and then vaporized it. I knew that fullerenes are likely to form in the carbon vapour produced from graphite even if $C_{60}$ grows into larger species. That would obscure any result, of course. To overcome this problem, I reduced the helium flow while keeping all other parameters constant. From my previous experiments, it was clear that fullerenes do not form in carbon vapour produced from graphite if the He pressure is not sufficiently high. Ablating the surface-coated $C_{60}$ graphite rod at reduced pressure resulted in a spectrum that showed $C_{60}$ abundant, but also remarkably the formation of a whole set of larger fullerenes (**Fig. 4**). The larger clusters were confirmed to be stable fullerene molecules by fragmentation studies. The "growth distribution" exactly mimicked the $C_{60}$ and larger distribution when fullerenes are from by vaporization of pure graphite.[1] Thus, identical growth processes occurred under both circumstances.

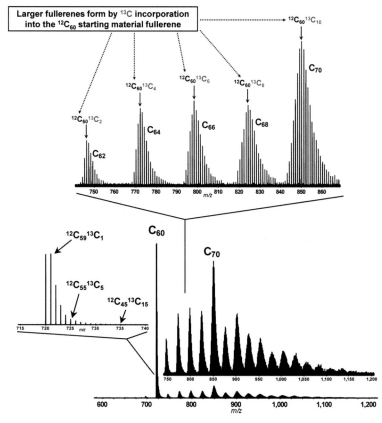

**Figure 5** Exposure of $C_{60}$ to pure $^{13}C$ vapour tracks growth. $C_{60}$ is clearly shown to grow bottom-up into larger fullerenes by incorporation of $C_2$ and C.

It needed to be proven that the larger fullerenes resulted from carbon incorporation into the pre-formed $C_{60}$. Therefore, I exposed $C_{60}$ to nearly pure $^{13}C$-carbon vapour under identical parameters to track carbon incorporation events (**Fig. 5**). This experiment neatly and unambiguously revealed that $C_2$ and atomic carbon were ingested by $C_{60}$ to grow into larger fullerenes. Facile atom exchange events were also observed. Harry contacted a former colleague and friend, Chris Ewels, at CNRS in France, who has significant expertise in the relevant theory and he kindly agreed to examine the processes we were observing theoretically and refine the newly elucidated carbon growth processes. Chris found that atom-exchange events could permit facile bond rearrangements, which provided a highly plausible explanation of how the highly symmetric icosahedral $C_{60}$ fullerene forms spontaneously during closed-cage growth from smaller cages. This also led us to further understanding and the development of a new technique for examining heterofullerene formation.[9]

The various open-network growth mechanisms which have been proposed find no support in the studies under our conditions, nor by our observations of the products obtained when metallofullerenes react with carbon vapour (**Fig. 6**). The metal atoms appear to be firmly locked inside the cages during growth, providing strong evidence that the cages remain essentially closed during

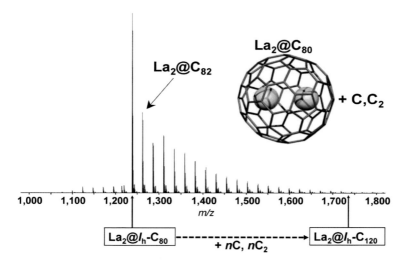

**Figure 6** The metallofullerene $La_2@C_{80}$ grows to larger species with both La atoms locked inside of the cage, demonstrating the cages remain closed.

the growth process. It takes considerable expertise together with special instrumentation to synthesize and isolate metallofullerenes which are not available commercially. The critical experiments we were able to carry out were made possible because Hisanori Shinohara promptly shipped samples of the compounds we needed from Japan. Collaboration was crucial at every step of the way. In addition, we showed that under certain conditions C$_{60}$ could grow in pure hydrogen. If the fullerene growth occurred by an open-network process, one would have expected it to be effectively disrupted by the presence of hydrogen. Thus all our observations support a fullerene a closed-network growth (CNG) mechanism very strongly.[10]

To put the final nail in the coffin of this long-standing problem, other fullerenes were exposed to carbon vapour and compared to the growth of C$_{60}$. The higher fullerenes C$_{70}$, C$_{76}$, C$_{84}$ were exposed individually to carbon vapour and the resulting growth patterns compared (**Fig. 7**). It was shown that under our conditions no C$_{60}$ formed by a deflation of larger fullerenes. Therefore, our results only make sense if C$_{60}$ forms through closed-network growth mechanism from smaller fullerenes. The observations also indicated that C$_{60}$ was the most resistant fullerene to carbon incorporation, and therefore, most resistant fullerene for growth into larger fullerenes. Consequently, the observation suggests that the dominance of C$_{60}$, over other fullerenes, generally observed during nucleation of carbon vapour may be due to a kind of "kinetic bottleneck" in closed-cage growth, thus providing an explanation to a long-standing carbon puzzle. The same growth trends are observed independent of the size of fullerene initially exposed to carbon vapour. The growth mechanism, therefore, does not seem to change with cage size and indicates that a single mechanistic path is dominant under our conditions. Very large nanostructures containing hundreds of atoms are formed after exposure

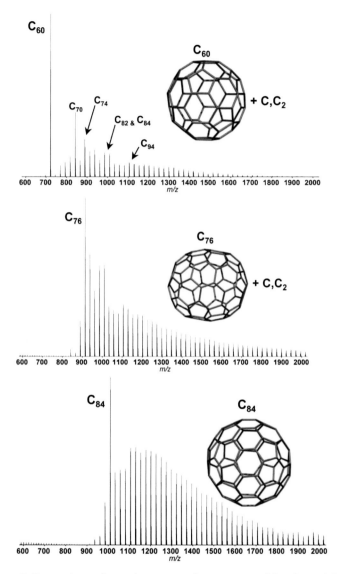

**Figure 7** Comparison of growth patterns after exposure of $C_{60}$, $C_{76}$ and $C_{78}$ to carbon vapour, under optimal growth conditions. The same growth trends are observed, indicating single mechanistic route. Importantly, absolutely no $C_{60}$ forms via the growth processes of larger fullerenes. Thus, $C_{60}$ forms by CNG from smaller fullerenes and is shown to be most resistant to carbon incorporation, further solving the riddle as to why $C_{60}$ forms as the most abundant fullerene.

of fullerenes to carbon vapour, so there appears to be no limit to the size of species that can be formed by the CNG mechanism. It is possible that some fraction of the larger structures are nanotubular.

## What It All Means

The understanding of how C$_{60}$ and other fullerenes, metallofullerenes and heterofullerenes form, elucidated by our work, should be useful in a broad range of areas. Researchers who work on fullerenes will, I hope, find our results of interest and useful in the development of their own experiments. The work, in fact, sheds new light on the processes that govern the self-assembly of carbon networks. The fundamental carbon chemistry revealed in our work is likely to be involved in the creation of nanostructures such as carbon nanotubes and possibly graphene, which form under similar conditions. In another recent chapter in the fullerene story, C$_{60}$ has, amazingly, been discovered to pop up in circumstellar shells, which suggests very strongly that it is a fundamental cosmic molecule.[11] Consequently, the CNG mechanism for fullerene formation should be of direct importance for understanding how fullerenes form throughout the universe as well, such as in the outflows of carbon stars and supernovae. It was a blast to have been able to contribute something to the C$_{60}$ story and work with some of the most amazing scientists.

I've learned that it is of utmost important to take full advantage of whatever opportunities come around; to take a chance, but be sure to give it your best shot and not be afraid of failure. It's interesting to contemplate the path I took to go from an inexperienced undergraduate researcher to one who feels he can take on whatever project comes his way, albeit still knowing that it will never be easy. Just like my first taste of collaborative research in Japan back in

2006, our success required an international team of people working together from different scientific backgrounds, cultures and parts of the world: Japan, Spain, UK, France and the USA, in this particular case. Ultimately, science brought us together as citizens not of a country, but rather as members of a planet with no borders that enabled us to do something we could not individually.

## References

1. H W Kroto, J R Heath, S C Obrien, R F Curl, R E Smalley, $C_{60}$: Buckminsterfullerene, *Nature*, **318**, 162–163 (1985).
2. H W Kroto, The stability of the fullerenes $C_{24}$, $C_{28}$, $C_{32}$, $C_{36}$, $C_{50}$, $C_{60}$ and $C_{70}$, *Nature*, **329**, 529–531 (1987).
3. A G Marshall, C L Hendrickson, G S Jackson, Fourier transform ion cyclotron resonance mass spectrometry: a primer, *Mass Spectrom Rev*, **17**, 1–35 (1998).
4. M A Duncan, Laser vaporization cluster sources, *Rev Sci Instrum*, 83 (2012). Invited review article.
5. H Shinohara, Endohedral metallofullerenes, *Rep Prog Phys*, **63**, 843–892 (2000).
6. T Guo et al., Uranium stabilization of $C_{28}$: a tetravalent fullerene, *Science*, **257**, 1661–1664 (1992).
7. P W Dunk et al., The smallest stable fullerene, $M@C_{28}$ (M = Ti, Zr, U): stabilization and growth from carbon vapor, *J Am Chem Soc*, **134**, 9380–9389 (2012).
8. E A Rohlfing, D M Cox, A Kaldor, Production and characterization of supersonic carbon cluster beams, *J Chem Phys*, **81**, 3322–3330 (1984).
9. P W Dunk et al., Formation of heterofullerenes by direct exposure of $C_{60}$ to boron vapor, *Angew Chem Int Ed*, **52**, 315–319 (2013).
10. P W Dunk et al., Closed network growth of fullerenes, *Nat Commun*, **3** (2012).
11. J Cami, J Bernard-Salas, E Peeters, S E Malek, Detection of $C_{60}$ and $C_{70}$ in a young planetary nebula, *Science*, **329**, 1180–1182 (2010).

# Part C

# References

1. H W Kroto, J R Heath, S C O'Brien, R F Curl and R E Smalley, *Nature*, **318**, 162–163 (1985).

2. W Krätschmer, L Lamb, K Fostiropoulos and D R Huffman, *Nature*, **347**, 354–358 (1990).

3. R Taylor, J P Hare, A K Abdul–Sada and H W Kroto, *J Chem Soc Chem, Commun*, 1423–1425 (1990).

4. H W Kroto, A W Allaf and S P Balm, *Chem Rev*, **91**, 1213–1235 (1991).

5. G Taubes, *Science*, **253**, 1476–1479 (1991).

6. E J Applewhite, *The Chemical Intelligence*, (July 1995), Vol 1 No. 3, Springer, http://www.4dsolutions.net/synergetica/eja1.html

7. H Hintenberger, J Franzen and K D Schuy, *Z Naturforsch,* **18a**, 1236–1237 (1963).

8. R Campargue, 24th International Symposium on Rarefied Gas Dynamics, AIP Conf. Proc. 762, 32–46 (2004).

9. R E Smalley, L Wharton and D H Levy, *J Chem Phys*, **63**, 4977–4989 (1975).

10. T G Dietz, M A Duncan, D E Powers and R E Smalley, *J Chem Phys*, **74**, 6511–6512 (1981).

11. A J Alexander, H W Kroto and D R M Walton, *J Mol Spectrosc*, **62**, 175–180 (1976).

12. H W Kroto, C Kirby, D R M Walton, L W Avery, N W Broten, J M MacLeod and T Oka, *Astrophysics J*, **219**, L133–L137 (1978).

13. E E Becklin, J A Frogel, A R Hyland, J Kristian, and G Neugebauer, *Astrophys J,* **158**, L133–L137 (1969).

14. A E Douglas, *Nature*, **269**, 130–132 (1977).

15. K L Day and D R Huffman, *Nature*, **243**, 50 (1973).

16. E A Rohlfing, D M Cox and A Kaldor, *J Chem Phys*, **81**, 3322–3330 (1984).

17. D E H Jones, *New Scientist*, **245**, 118–119 (1966) [Daedalus column].

18. E Osawa, *Kagaku*, **25**, 854–863 (1970) [in Japanese].

19. J R Heath, S C O'Brien, Q Zhang, Y Liu, R F Curl, H W Kroto, F K Tittel and R E Smalley, *J Am Chem Soc*, **107**, 7779–7780 (1985).

20. H W Kroto, *Nature*, **329**, 529–531 (1987).

21. T G Schmalz, W A Seitz, D J Klein and G E Hite, *J Am Chem Soc*, **110**, 1113–1127 (1988).

22. F D Weiss, J L Elkind, S C O'Brien, R F Curl and R E Smalley, *J Am Chem Soc*, **110**, 4464–4465 (1988).

23. H W Kroto and K G McKay, *Nature*, **331**, 328–331 (1988).

24. D M Cox, D J Trevor, K C Reichmann and A Kaldor, *J Am Chem Soc*, **108**, 2457–2458 (1986).

25. W Krätschmer, K Fostiropoulos and D R Huffman, in *Dusty Objects in the Universe*, ed. E Bussoletti and A A Vittone, Kluwer, 89–93 (1989).

26. P W Dunk, N K Kaiser, M Mulet-Gas, A Rodríguez-Fortea, J M Poblet, H Shinohara, C L Hendrickson, A G Marshall, and H W Kroto, *J Am Chem Soc*, **134**, 9380–9389 (2012).

27. R C Haddon, A F Hebard, M J Rosseinsky, D W Murphy, S J Duclos, K B Lyons, B Miller, J M Rosamilia, R M Fleming, A Kortan, S H Glarum, A V Makhija, A J Muller, R H Eick, S M Zahurak, R Tycko, G Dabbagh, F A Thiel, *Nature*, **350**, 320–322 (1991).

28. G von Helden, M-T Hsu, P R Kemper and M T Bowers, *J Chem Phys*, **95**, 3835–3837 (1991).

29. J R Heath, in *Fullerenes: Synthesis, Properties and Chemistry of Large Carbon Clusters*, ACS Symposium Series No. 481, ed. G S Hammond and V J Kuck, 1–23 (1992).

30. M Endo and H W Kroto, *J Phys Chem*, **96**, 6941–6944 (1992).

31. P W Dunk, N K Kaiser, C L Hendrickson, J P Quinn, C P Ewels, Y Nakanishi, Y Sasaki, H Shinohara, A G Marshall and H W Kroto, *Nat Commun*, **3**, 855 (2012).

32. H W Kroto and M Jura, *Astron Astrophys*, **263**, 275–280 (1992).

33. J Cami, J Bernard-Salas, E Peeters and S E Malek, *Science*, **329**, 1180 (2010).

# Index

acetone 64
aluminum 64–65
arc-discharged carbon 25
argon ion laser 43
aromaticity
  2D 99
  3D 89–90, 98
  planar 89
  spherical 90
astronomy 113, 125, 127, 131
atoms
  chlorine 107–108
  titanium 161–162

benzene 28, 61, 102–104, 125, 138–141
benzene rings 103–104
biphenyl 103–104
boranes 89–90
buckminsterfullerene 4, 22–23, 80, 115
buckminsterfullerene structure 19, 22
buckyballs 75, 79

$C_{60}$
  chemical synthesis of 101–108
  discovery of 6, 8, 10, 28, 67, 71, 91, 101, 106, 169
  synthesizing 105, 107–108

$C_{60}$ buckminsterfullerene 87
$C_{60}$ buckminsterfullerene formation process 149–174

carbon 7, 10, 47–48, 75, 80, 83, 90, 92, 115, 131, 133, 139, 162–163, 166, 168
carbon atoms 7, 15, 17, 19, 25, 29, 47–48, 143, 154, 160, 162
  pyramidalized trigonal 102
carbon chains 10, 12, 75, 83
carbon chains in space and stars 6, 9
carbon chemistry 28, 82
carbon cluster experiments 155
carbon cluster source 156–157
carbon cluster studies 6–7
carbon clusters 63, 75, 82, 154, 156, 158, 169
carbon dust 119
carbon in space dust 6, 10
carbon materials science 28
carbon nanotubes 92–93, 108, 150, 174
carbon particles 114, 116
carbon rod 161–162
carbon vapour 29, 164, 166–174
closed-network growth (CNG) 172–174
CNG, see closed-network growth

corannulene  98–99, 105–106
  synthesis of  106
cyanopolyynes  18–19

decacyclene  107
DIBs, *see* diffuse interstellar bands
diffuse interstellar bands (DIBs)
  6, 10–12, 31
dye laser  57, 59–61

electron beam  56
electrons  97–98, 163
excimer laser  63, 65–67

fast flow cluster reactor  84
flash vacuum pyrolysis (FVP)  103,
  105–108
formaldehyde  54, 59–60
FT-ICR-MS  28–30
fullerene addition reactivity
    pattern  28
fullerene arc apparatus  128
fullerene $C_{28}$  160
  endohedral  161
  small  28
fullerene cages  164
  concentric  21
fullerene chemistry  75, 82
fullerene collaborations  125
fullerene conjecture  24
fullerene creation process  164
fullerene dynamics  23
fullerene experiments  58
fullerene formation  163–164,
    166, 174

fullerene growth  165, 172
fullerene research  84, 93
  post-1990  29
fullerene science  92–93, 125,
    140–141, 143
  birth of  140–143
fullerene soot generator  144
fullerene synthesis  108, 157
fullerenes
  closed-cage  29
  empty cage  165
  endohedral  143
  giant  21, 103
  smallest  154, 157, 160–161
  surface-coated  166
FVP, *see* flash vacuum pyrolysis

GaAs  74–75, 79
geodesic curvature  104–105
geodesic polyarenes  105–107
graphite  17, 81, 103, 130, 166,
    169
  pure  168–169
graphite rod  166–167, 169

Harry Kroto's Nobel Prize  146
heterofullerene  108, 171
heterofullerenes  108, 174
higher fullerenes  107
  synthesis of  108
hollow carbon molecule  48
hydrocarbon flames  30
hydrocarbons  102–103
hydrogen  11, 41, 103, 105, 107,
    154, 160, 172
  atoms  103, 106

IMPR, *see* isolated multiplet pentagon rule

intramolecular vibrational relaxation (IVR) 62

ionization 55, 61, 63

IPR, *see* isolated pentagon rule

iron 64, 81

iron atoms 64

isolated multiplet pentagon rule (IMPR) 19–20

isolated pentagon rule (IPR) 6, 19–20

isotope separation, uranium 44

IVR, *see* intramolecular vibrational relaxation


Kamel soot 115, 117, 119–120, 122, 124

Kamel spectrum 115, 118


laser ablation 166–167

laser-induced fluorescence (LIF) 61, 67

laser pyrolysis 50–51

laser vaporization source 61, 65

LIF, *see* laser-induced fluorescence


metallofullerene 163

metallofullerene $La_2@C_{80}$ 171

metallofullerenes 81, 160, 162, 164, 171–172, 174

metastable fullerene cations 83

molecular beams 39–40, 42–43, 54–55, 59, 64, 80


nanotubes 28–29, 92

nitrogen 11, 29, 41, 43, 54

nude ionization gauge 58–59


PAHs
  *see* polycyclic aromatic hydrocarbons

planar 105–107

phenyl radical 63

photoionization 60, 65

photolysis 63–64

photolysis laser 64

polycyclic aromatic hydrocarbons (PAHs) 104, 107


ruby laser 50


semiconductor clusters 67, 74, 79

superaromaticity 98–99

superphane 102


Td-$C_{28}$ fullerene 161

Td-$C_{28}$ fullerene isomer 154

thermal cyclodehydrogenations 105, 107

titanium 28, 161–162

toluene 61, 125, 138, 140

Two Micron Sky Survey 69–70


uranium complexes 62

vaporization laser 65

Printed and bound by CPI Group (UK) Ltd, Croydon, CR0 4YY

23/10/2024

01777708-0001